Intermediate 1

Mathematics

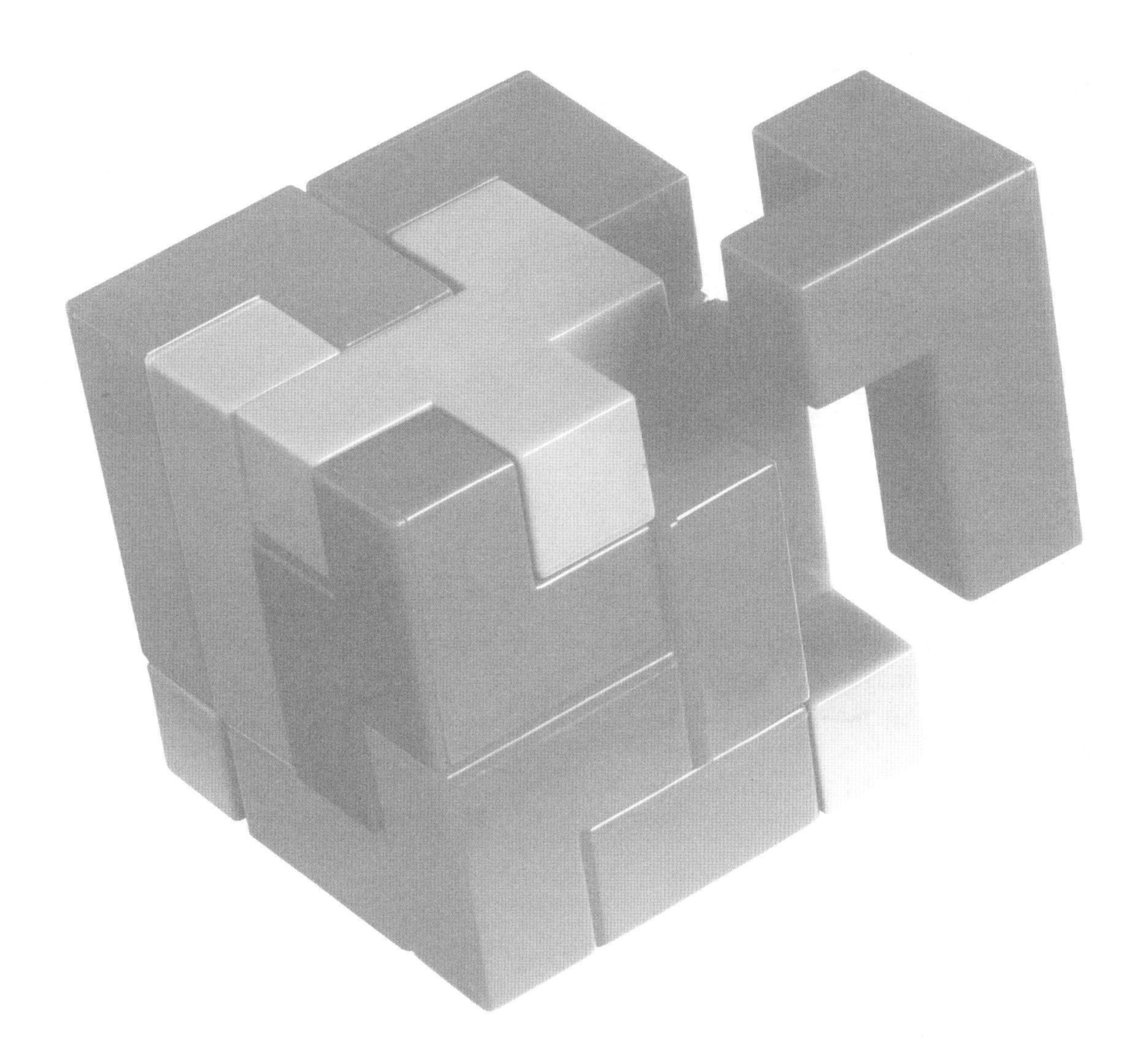

Ken Nisbet

Contents

Introduction

Unit 1

Unit 2

Unit 3

Unit 4

Answers

Index

This book, your exam and your calculator

Using this book

About this book

This Success Guide will help you refresh your memory of the Intermediate 1 Mathematics course, so that you can succeed in the Intermediate 1 Mathematics Unit 'NAB' exams. It can also help you prepare for the end-of-course exam. This book follows the Unit structure of the course specified in the National Course Specification Document.

However, you don't become good at mathematics just by reading books – although they will give you the knowledge and skills you need to do so. The more you practise, the better you become – so use this book to start you off, then get out there and start problem-solving!

Top tips

Tips on mathematics and exam techniques are throughout the book: make sure you read and learn them.

Quick tests

These are at the end of each topic. If you have difficulty with their questions after revising the topic, then you should practice that topic more.

Top Tip
Your teacher wants you to pass, and will be happy to answer any of your questions.

Top Tip
Use the Quick Tests to identify your strengths and weaknesses.

Exam practice

The best practice for tests and exams is sitting practice papers under 'exam conditions' at home: no TV, radio, PC or mobile phone! Leckie & Leckie will soon publish a wide range of practice exam papers – see www.leckieandleckie.co.uk for more details. These books will have fully worked answers for you to assess your exam skills. You can also download last year's actual exams from www.sqa.org.uk.

Top Tip
When you are revising maths, more than half of your time should be spent working out questions.

Getting more help

You can always go back to your textbook, your notes or your teacher for more examples and explanations if there's anything you're not sure about. If you are learning on your own, you may need to find a knowledgeable friend to help you out occasionally.

The course structure and the exam

The course structure is:

mandatory	Mathematics 1 (Intermediate 1)	–	Unit 1 of this guide
	Mathematics 2 (Intermediate 1)	–	Unit 2 of this guide
optional	Mathematics 3 (Intermediate 1)	–	Unit 3 of this guide
	Applications of Mathematics (Intermediate 1)	–	Unit 4 of this guide

Units 3 and 4 are optional in that you will study one of them, not both. If you intend to study mathematics beyond Intermediate 1 level, you are expected to study Mathematics 3, not Applications of Mathematics.

You have to pass a Unit test in each Unit you study to gain a course award, which then allows you to take the end-of-course exam. However, it is very important to remember that the Unit tests contain only the easier bits of the course, and that their questions are very predictable, so you could get very good at Unit test questions without really being close to passing the exam.

You must make sure you sit the right version of the exam: either Mathematics 1, 2 and 3 or Mathematics 1 and 2 and Applications of Mathematics. Both version have two exam papers.

Your calculator

Whichever version of the exam you sit, you can't use a calculator in Paper 1 (the first, and shorter, paper) but you will need to use a scientific calculator in Paper 2.

Some guidance is given in the text on when and how to use your calculator, and when **not** to.

Top Tip

Be sure to bring your own calculator to Paper 2, and make sure you know how to use it.

Rounding Numbers

Approximations

Number answers to problems don't always need to be given exactly. There are rules for giving the correct approximate answers. Here are some examples:

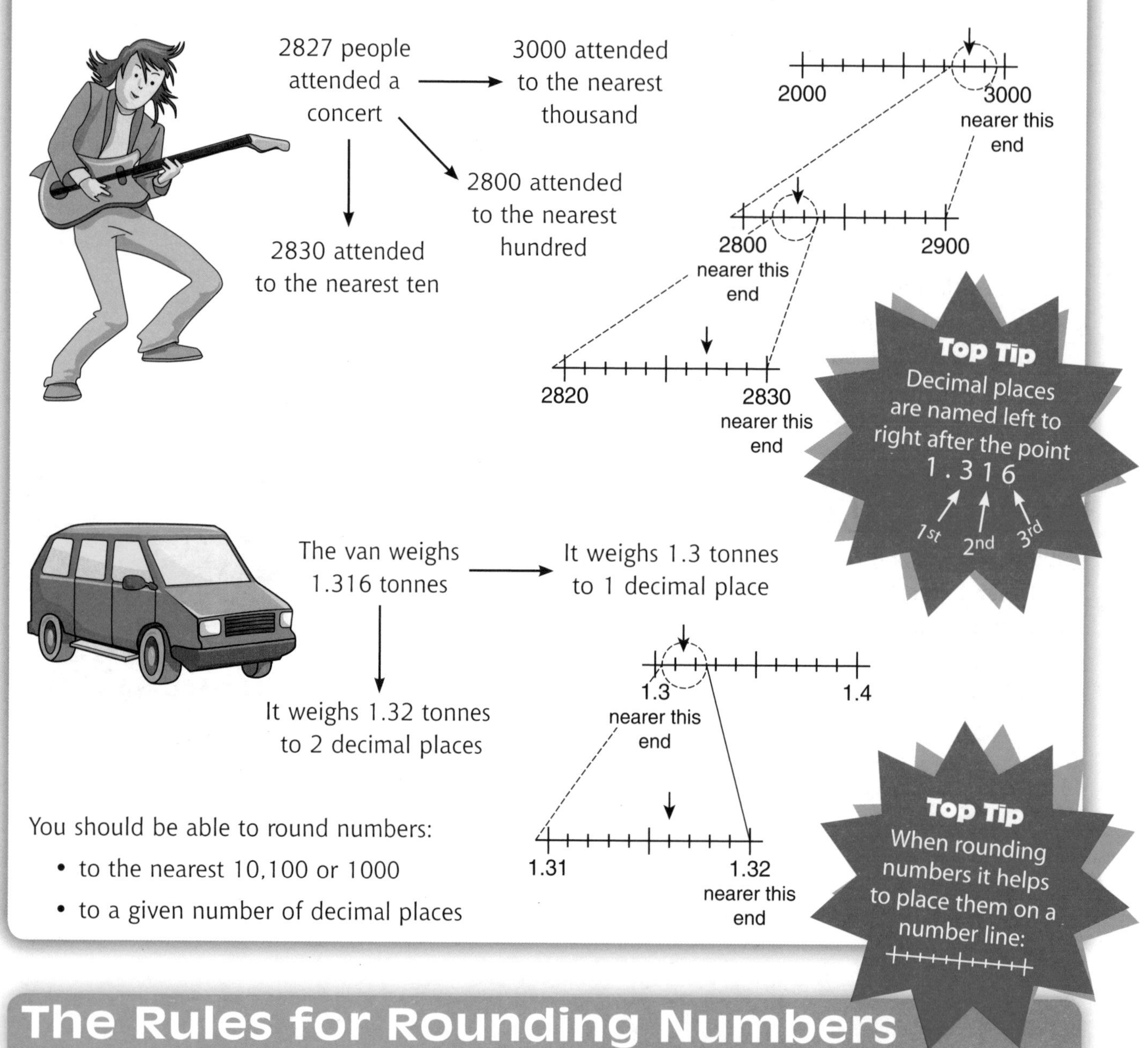

You should be able to round numbers:

- to the nearest 10,100 or 1000
- to a given number of decimal places

The Rules for Rounding Numbers

Last Digit stays the same

If the next digit is 0, 1, 2, 3 or 4 the last digit stays the same

26.4 3 8 becomes 26.4 to 1 decimal place

(4: leave the same; 3: next digit is 3)

Last Digit goes up 1

If the next digit is 5, 6, 7, 8 or 9 the last digit goes up 1

2 8 8 becomes 290 to the nearest ten.

(8: goes up 1; 8: next digit is 8)

Rounding and Units

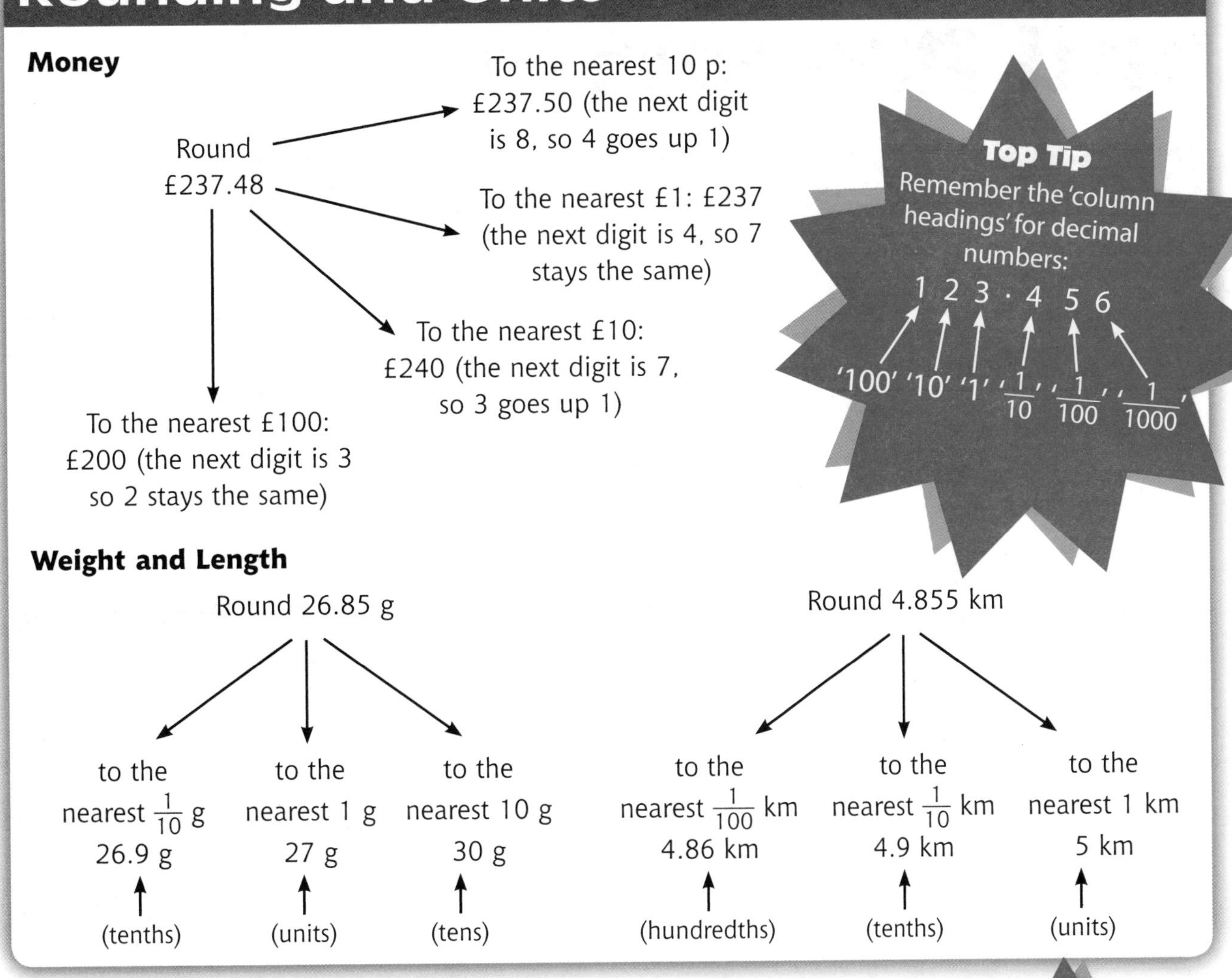

In Context

Sometimes the 'Rules for Rounding numbers' don't apply. Suppose you need 13 AA Batteries and they are sold in packs of 4. How many packs do you need? Dividing 13 by 4 gives 3.25 but 3 packs of 4 is not enough. You need to buy 4 packs. In this case 3.25 becomes 4.

Top Tip

Remember fractions to decimals:

eg $\frac{8}{1000} = 0.008$

$\frac{3}{100} = 0.03$

Quick Test

1. Write 5.26 m^2 correct to 1 decimal place.
2. Round 16 675 cm^3 to the nearest thousand cubic centimeters.
3. Round off 12 445 dollars to the nearest 10 dollars.
4. A printer needs 13 600 sheets of paper. He buys them in packs of 500 sheets. How many packs should he buy?

Top Tip

Most money answers should be written to the nearest penny. This is £26.14.

26.13941

Answers: 1. 5.3 m^2 **2.** 17 000 cm^3 **3.** 12 450 dollars **4.** 28 packs

Percentages, Fractions and Proportion

Percentages to Fractions

Remember 'percent' means 'per hundred'. For example 40% = $\frac{40}{100} = \frac{4}{10} = \frac{2}{5}$

There are some common percentages you should know:

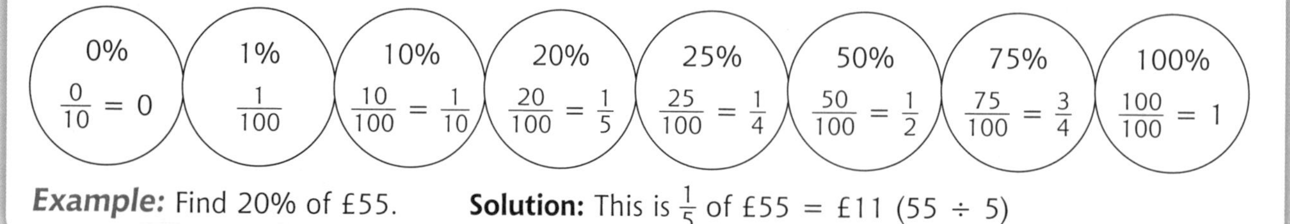

Example: Find 20% of £55.

Solution: This is $\frac{1}{5}$ of £55 = £11 (55 ÷ 5)

Finding a Percentage of a Quantity

Step 1: Divide the Quantity by 100 (to find 1%)

Step 2: Multiply by the given percentage

Example: Find 8% of £127

Solution: 127 ÷ 100 × 8
= 1.27 × 8
= £10.16

Example: Find $2\frac{1}{2}$% of £6.60

Solution: 6.6 ÷ 100 × 2.5
= 0.066 × 2.5 = 0.165
= 17p (to the nearest penny)

Top Tip
To change a fraction to a % multiply by 100%
$\frac{1}{4} = \frac{1}{4} \times 100\% = 25\%$
$\frac{3}{5} = \frac{3}{5} \times 100\% = 60\%$

Discounts

Discounts are taken off a price – this is a subtraction

This £350 Laptop is being offered with a 5% discount. What is the price after the discount?

Solution:
5% of £350
= 350 ÷ 100 × 5
= 3.5 × 5
= 17.5
The discount is £17.50
The price is £350 – £17.50 = £332.50

Top Tip
In Sales you get Discounts – remember to subtract. VAT is an extra added to your bill.

VAT

VAT stands for 'Value Added Tax' and will be added to your bill – this is an addition. If the VAT rate is 15% and your bill is £100 then £15.00 (15% of £100) will be added to your bill. You will be charged £115.

A Garage bill is £48 with VAT charged at 15%. Here is the calculation:
VAT = 15% of £48 = 48 ÷ 100 × 15 = £7.20. You will pay £48 + £7.20 = £55.20.

Simple Interest

If you keep money in a bank (savings or investments) the bank will pay you. This extra money is called interest and is calculated using a Rate of Interest given for a year (per annum or p.a.)

Example: Iain invests £1200 at 4% per annum. Calculate the interest he should receive.

Solution: 4% of £1200 = 1200 ÷ 100 × 4 = 12 × 4 = £48

He receives £48 interest for 1 year.

Note: For 3 months ($\frac{1}{4}$ of a year) his interest is 48 ÷ 4 = £12

For 4 months ($\frac{1}{3}$ of a year) his interest is 48 ÷ 3 = £16

Top Tip

Learn these fractions of a year:

1 year = 12 months

$\frac{1}{2}$ of a year = 6 months

$\frac{1}{3}$ of a year = 4 months

$\frac{2}{3}$ of a year = 8 months

$\frac{1}{4}$ of a year = 3 months

$\frac{3}{4}$ of a year = 9 months

One Quantity as a % of Another Quantity

Example: £20 is increased by £2 to £22. What percentage increase is this?

Solution: Percentage increase $= \frac{\text{increase}}{\text{original amount}} \times 100\%$

$= \frac{2}{20} \times 100\% = 10\%$ So this is a 10% increase.

Note: £2 is 10% of £20.

Example: £25 is decreased by £5 to £20. What percentage decrease is this?

Solution: Percentage decrease $= \frac{\text{decrease}}{\text{original amount}} \times 100\%$

$= \frac{5}{25} \times 100\% = 20\%$ So this is a 20% decrease.

Direct Proportion

If 8 cans of juice cost £2.40 then 1 can costs 30p (£2.40 ÷ 8)

So 5 cans will cost £1.50 (30p × 5). This is an example of DIRECT PROPORTION.

Example: The number of rolls of wallpaper needed is proportional to the area of the wall to be decorated. 7 rolls cover 175 m^2. What area would 12 rolls cover?

Solution: 1 roll covers 175 ÷ 7 = 25 m^2. So 12 rolls cover 25 × 12 = 300 m^2.

Quick Test

1. Zoe paid £80 for a skirt then sold it for £100. Express the profit she made as a percentage of what she paid for the skirt.
2. Calculate 30% of £240.
3. £36 is increased by £9. What percentage increase is this?
4. Calculate $\frac{4}{9}$ of £27.
5. **(a)** Mia invested £580 at 3% per annum. How much interest does she get
   (a) for 1 year? **(b)** for 3 months? **(c)** for 8 months?
6. What do you pay for a £28.80 bill if VAT is added at $17\frac{1}{2}$%?

Top Tip

To find $\frac{3}{5}$ of £80 find $\frac{1}{5}$ first by dividing by 5

80 ÷ 5 = 16

So $\frac{1}{5}$ of £80 = £16

and $\frac{3}{5}$ of £80 = 16 × 3 = £48.

Divide by the bottom number. Multiply by the top

Answers: **1.** 25% **2.** £72.00 **3.** 25% **4.** £12 **5. (a)** £17.40 **(b)** £4.35 **(c)** £11.60 **6.** £33.84

Areas

Units

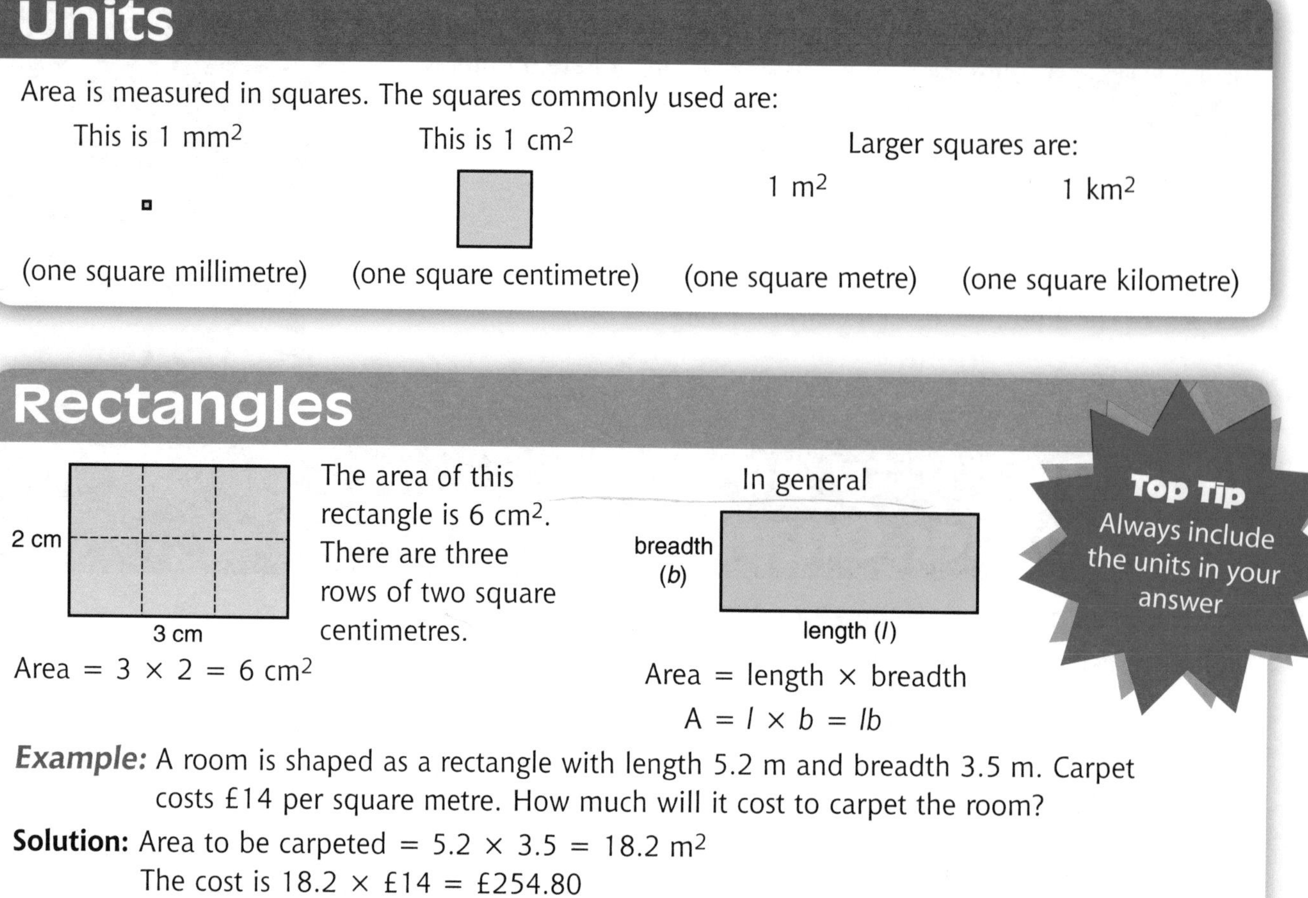

Area is measured in squares. The squares commonly used are:

This is 1 mm^2 (one square millimetre)

This is 1 cm^2 (one square centimetre)

Larger squares are:

1 m^2 (one square metre)

1 km^2 (one square kilometre)

Rectangles

The area of this rectangle is 6 cm^2. There are three rows of two square centimetres.

Area = $3 \times 2 = 6$ cm^2

In general

Area = length × breadth

$A = l \times b = lb$

Top Tip
Always include the units in your answer

Example: A room is shaped as a rectangle with length 5.2 m and breadth 3.5 m. Carpet costs £14 per square metre. How much will it cost to carpet the room?

Solution: Area to be carpeted = $5.2 \times 3.5 = 18.2$ m^2

The cost is $18.2 \times$ £14 = £254.80

Triangles

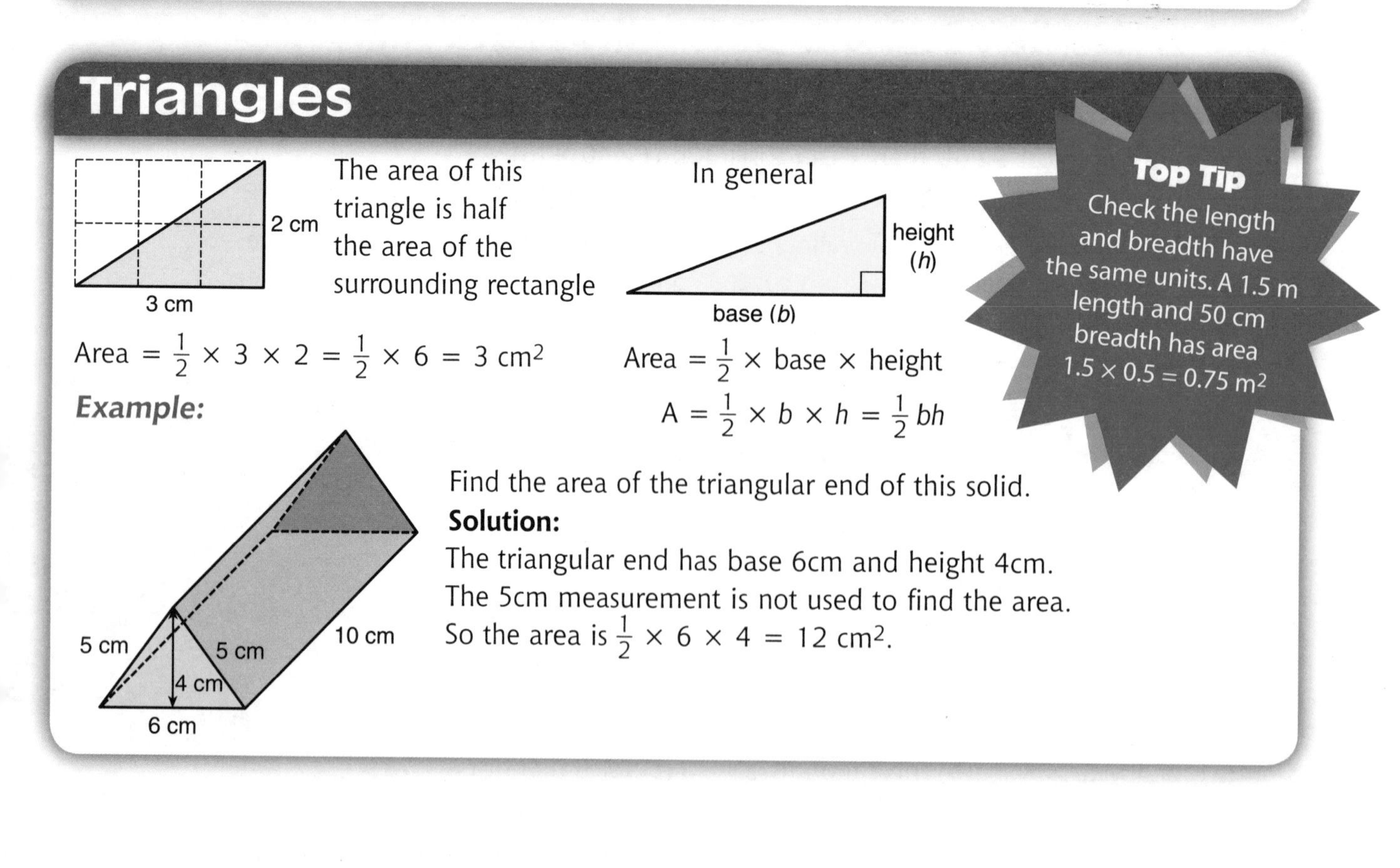

The area of this triangle is half the area of the surrounding rectangle

Area = $\frac{1}{2} \times 3 \times 2 = \frac{1}{2} \times 6 = 3$ cm^2

In general

Area = $\frac{1}{2}$ × base × height

$A = \frac{1}{2} \times b \times h = \frac{1}{2} bh$

Top Tip
Check the length and breadth have the same units. A 1.5 m length and 50 cm breadth has area $1.5 \times 0.5 = 0.75$ m^2

Example:

Find the area of the triangular end of this solid.

Solution:

The triangular end has base 6cm and height 4cm.

The 5cm measurement is not used to find the area.

So the area is $\frac{1}{2} \times 6 \times 4 = 12$ cm^2.

Composite Shapes

Shapes can be made up of several triangles and rectangles. You should split the shape into separate rectangles and triangles. Find the area of each of these pieces and then add these areas to find the total area of the larger shape.

Example: Find the area of the kitchen worktop shown in this diagram

Solution: Divide the shape into two rectangles as shown.

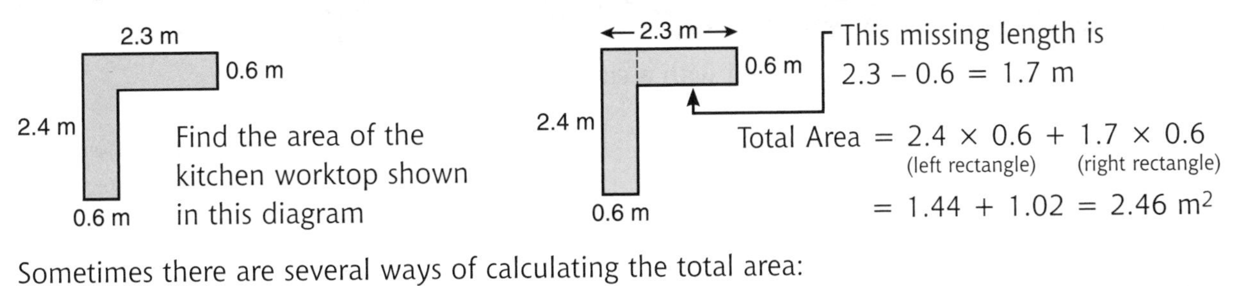

This missing length is $2.3 - 0.6 = 1.7$ m

Total Area $= 2.4 \times 0.6 + 1.7 \times 0.6$
(left rectangle) (right rectangle)

$= 1.44 + 1.02 = 2.46\ m^2$

Sometimes there are several ways of calculating the total area:

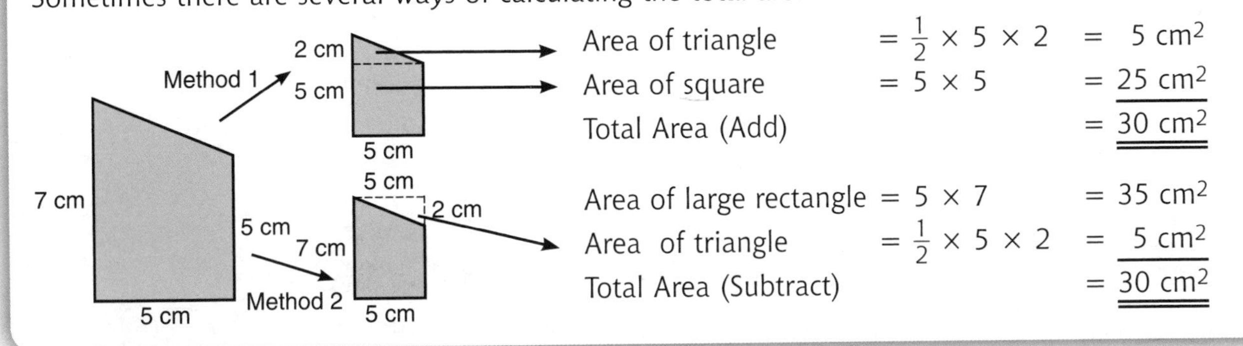

Area of triangle $= \frac{1}{2} \times 5 \times 2 = 5\ cm^2$
Area of square $= 5 \times 5 = 25\ cm^2$
Total Area (Add) $= 30\ cm^2$

Area of large rectangle $= 5 \times 7 = 35\ cm^2$
Area of triangle $= \frac{1}{2} \times 5 \times 2 = 5\ cm^2$
Total Area (Subtract) $= 30\ cm^2$

Cost Problems

You may be asked to calculate the cost of, for instance, carpeting an L-shaped room, painting a wall with a door in it or gravelling a path round a garden. You will need to calculate the total area of the shape involved and then multiply it by the cost of 1 m^2 which you will be given.

Example: The rectangular 8m × 5m grass area of a garden is surrounded by a path that is 0.5m wide. It costs £5 per square metre to gravel the path. What is the total cost to gravel the path?

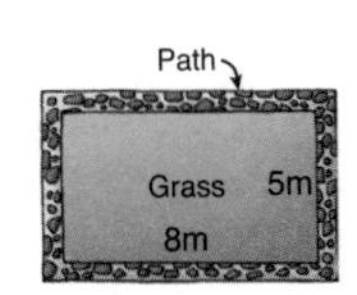

Solution: Length of outside rectangle $= 8 + 2 \times 0.5 = 9$ m

Breadth of outside rectangle $= 5 + 2 \times 0.5 = 6$ m

Area of path $= 9 \times 6 - 8 \times 5$
(outside rectangle) (inside rectangle)

$= 54 - 40 = 14\ m^2$

Cost $= 14 \times £5 = £70$.

Quick Test

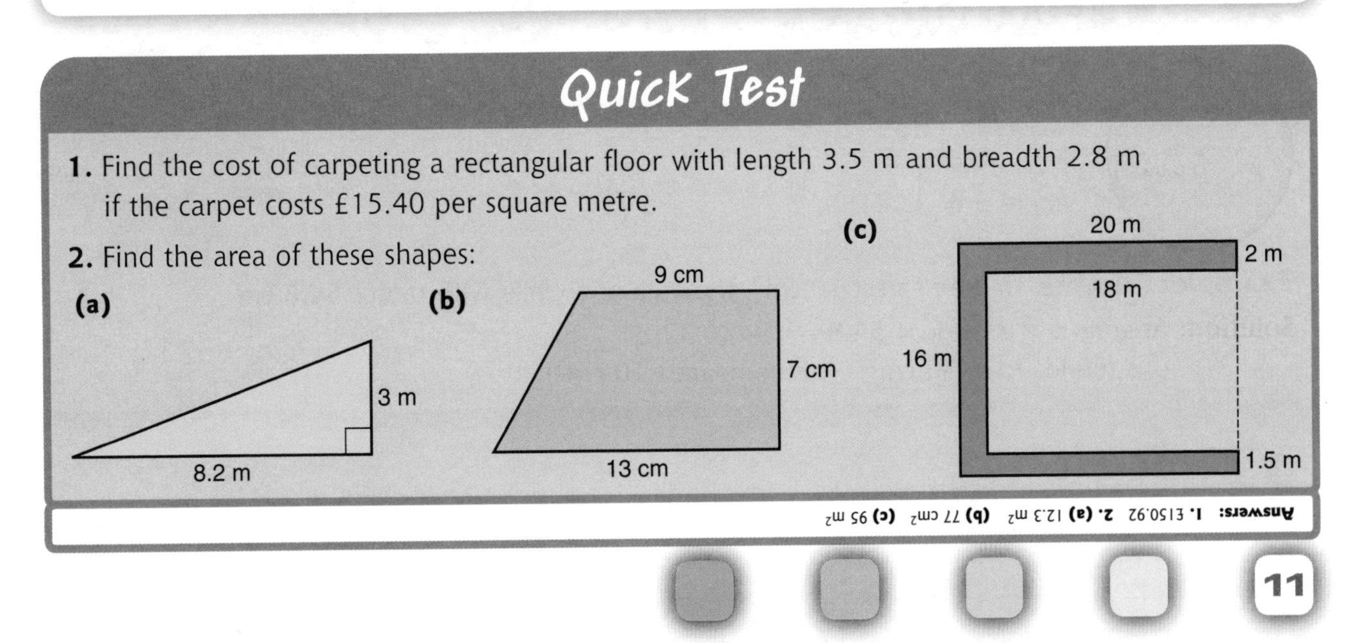

1. Find the cost of carpeting a rectangular floor with length 3.5 m and breadth 2.8 m if the carpet costs £15.40 per square metre.
2. Find the area of these shapes:

(a) **(b)** **(c)**

Answers: 1. £150.92 **2. (a)** 12.3 m² **(b)** 77 cm² **(c)** 95 m²

Circles and Volumes of Cuboids

Basic Facts

Here are the names and letters associated with a circle:

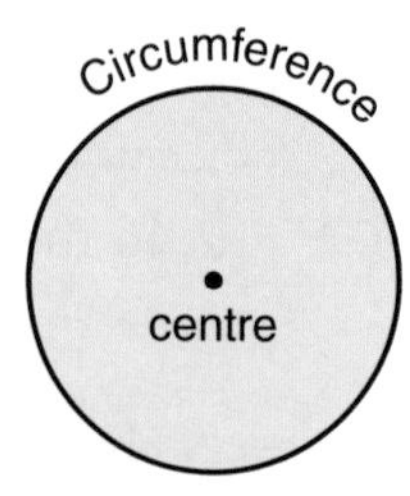

Use C for the length of the Circumference

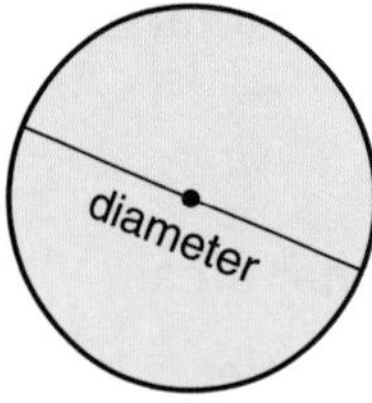

Use d for the length of the diameter

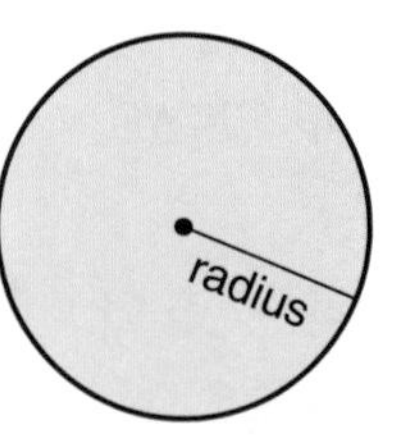

Use r for the length of the radius

Notice that the diameter is twice the length of the radius $d = 2 \times r$

Top Tip
The formulae
$C = \pi d$
$A = \pi r^2$ are given in the Exam but you should learn them.

Circumference Formula

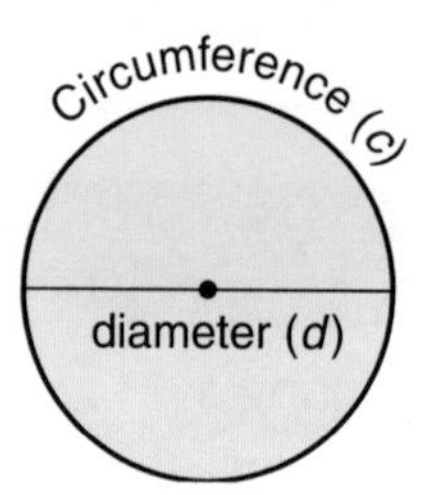

Circumference = $\pi \times$ diameter

$c = \pi \times d$

where $\pi = 3.14159...$

Only use 3.14 for π if the question tells you to do this otherwise your answer may not be accurate.

π is 'pi' a letter from the Greek alphabet

The π button on your calculator gives it's value.

Example: Calculate, to 1 decimal place, the circumference of a circle with radius 4.1 cm.

Solution: First double the radius to find the length of the diameter:

diameter $= 2 \times r = 2 \times 4.1 = 8.2$ cm

So Circumference $= \pi \times$ diameter $= \pi \times 8.2 = 25.7610.$

This is rounded to 25.8 cm (to 1 decimal place)

Area Formula

Area = $\pi \times$ radius $\times$ radius

$A = \pi \times r \times r$

$A = \pi \times r^2$

where $\pi = 3.14159...$

Top Tip
In the calculator allowed paper always use the π button and round your final answer.

Example: Calculate, to the nearest 10 cm^2, the area of a circle with radius 54.8 cm.

Solution: Area $= \pi \times r^2 = \pi \times 54.8^2 = 9434.32...$

this rounds to 9430 cm^2 (to the nearest 10 cm^2)

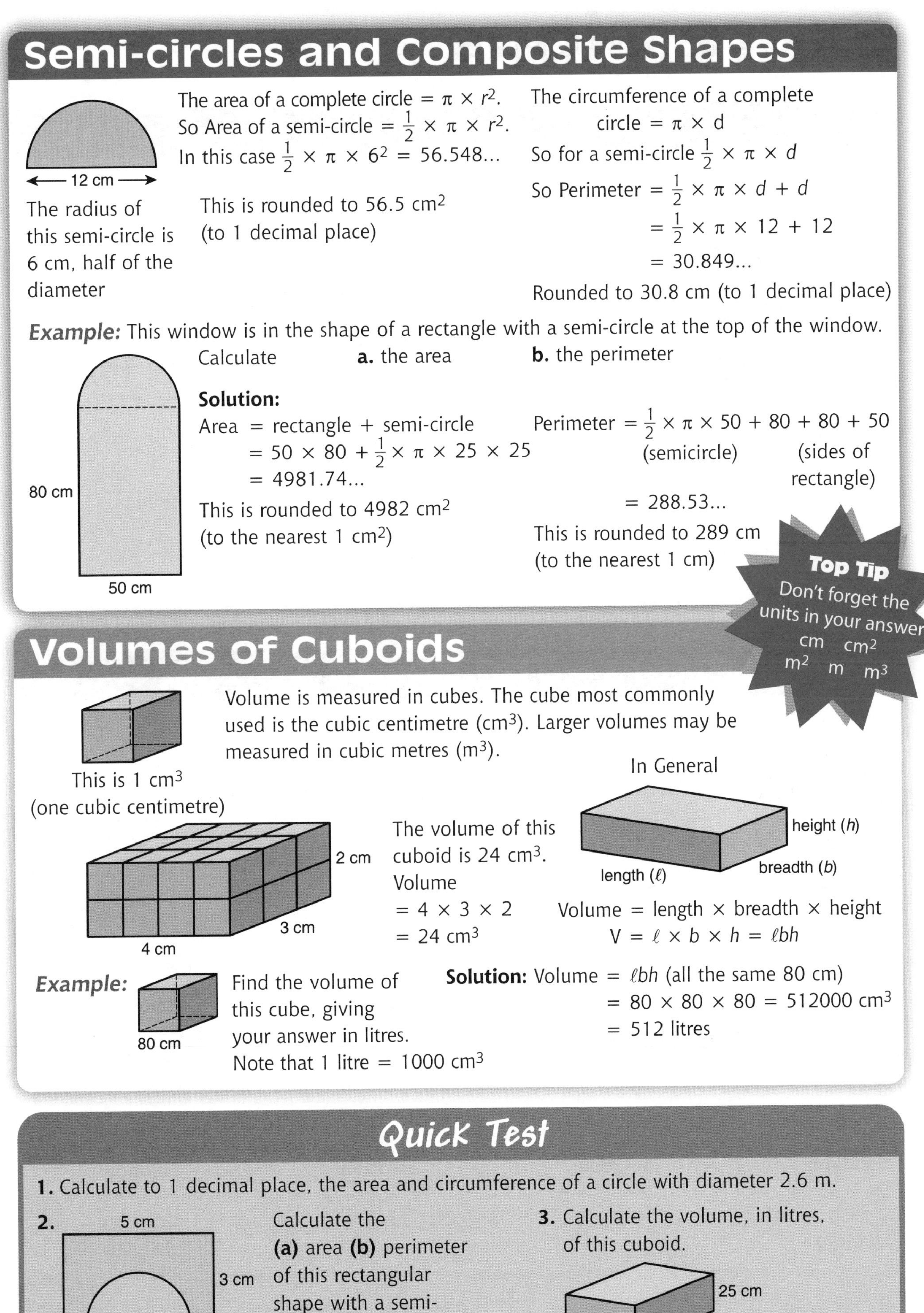

Semi-circles and Composite Shapes

The radius of this semi-circle is 6 cm, half of the diameter

The area of a complete circle $= \pi \times r^2$.
So Area of a semi-circle $= \frac{1}{2} \times \pi \times r^2$.
In this case $\frac{1}{2} \times \pi \times 6^2 = 56.548...$

This is rounded to 56.5 cm^2 (to 1 decimal place)

The circumference of a complete circle $= \pi \times d$
So for a semi-circle $\frac{1}{2} \times \pi \times d$
So Perimeter $= \frac{1}{2} \times \pi \times d + d$
$= \frac{1}{2} \times \pi \times 12 + 12$
$= 30.849...$
Rounded to 30.8 cm (to 1 decimal place)

Example: This window is in the shape of a rectangle with a semi-circle at the top of the window.
Calculate **a.** the area **b.** the perimeter

Solution:
Area = rectangle + semi-circle
$= 50 \times 80 + \frac{1}{2} \times \pi \times 25 \times 25$
$= 4981.74...$
This is rounded to 4982 cm^2 (to the nearest 1 cm^2)

Perimeter $= \frac{1}{2} \times \pi \times 50 + 80 + 80 + 50$
(semicircle) (sides of rectangle)
$= 288.53...$
This is rounded to 289 cm (to the nearest 1 cm)

Top Tip
Don't forget the units in your answer
cm cm^2 m^2 m m^3

Volumes of Cuboids

Volume is measured in cubes. The cube most commonly used is the cubic centimetre (cm^3). Larger volumes may be measured in cubic metres (m^3).

This is 1 cm^3 (one cubic centimetre)

The volume of this cuboid is 24 cm^3.
Volume
$= 4 \times 3 \times 2$
$= 24$ cm^3

In General

Volume = length × breadth × height
$V = \ell \times b \times h = \ell bh$

Example: Find the volume of this cube, giving your answer in litres.
Note that 1 litre = 1000 cm^3

Solution: Volume $= \ell bh$ (all the same 80 cm)
$= 80 \times 80 \times 80 = 512000$ cm^3
$= 512$ litres

Quick Test

1. Calculate to 1 decimal place, the area and circumference of a circle with diameter 2.6 m.

2. Calculate the **(a)** area **(b)** perimeter of this rectangular shape with a semi-circle removed.

3. Calculate the volume, in litres, of this cuboid.

Answers: 1. A = 5.3 m^2, C = 8.2 m **2. (a)** 11.5 cm^2 **(b)** 17.7 cm **3.** 75 litres

Expressions

Order of Operations

Your calculator should give 11 as the answer to $3 + 2 \times 4$. Notice it has calculated 2×4 first then $3 + 8$ to give 11. Here is the order you should work calculations:

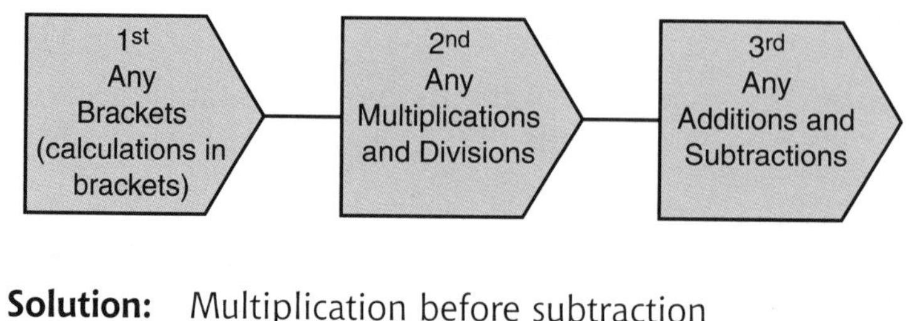

Examples: $340 - 15 \times 8$ — **Solution:** Multiplication before subtraction
$340 - 120 = 220$

$(2 + 9) \times 4$ — **Solution:** Bracket calculation before multiplication
$11 \times 4 = 44$

$2 \times 3 + 4 \times 5$ — **Solution:** Multiplications before the addition
$6 + 20 = 26$

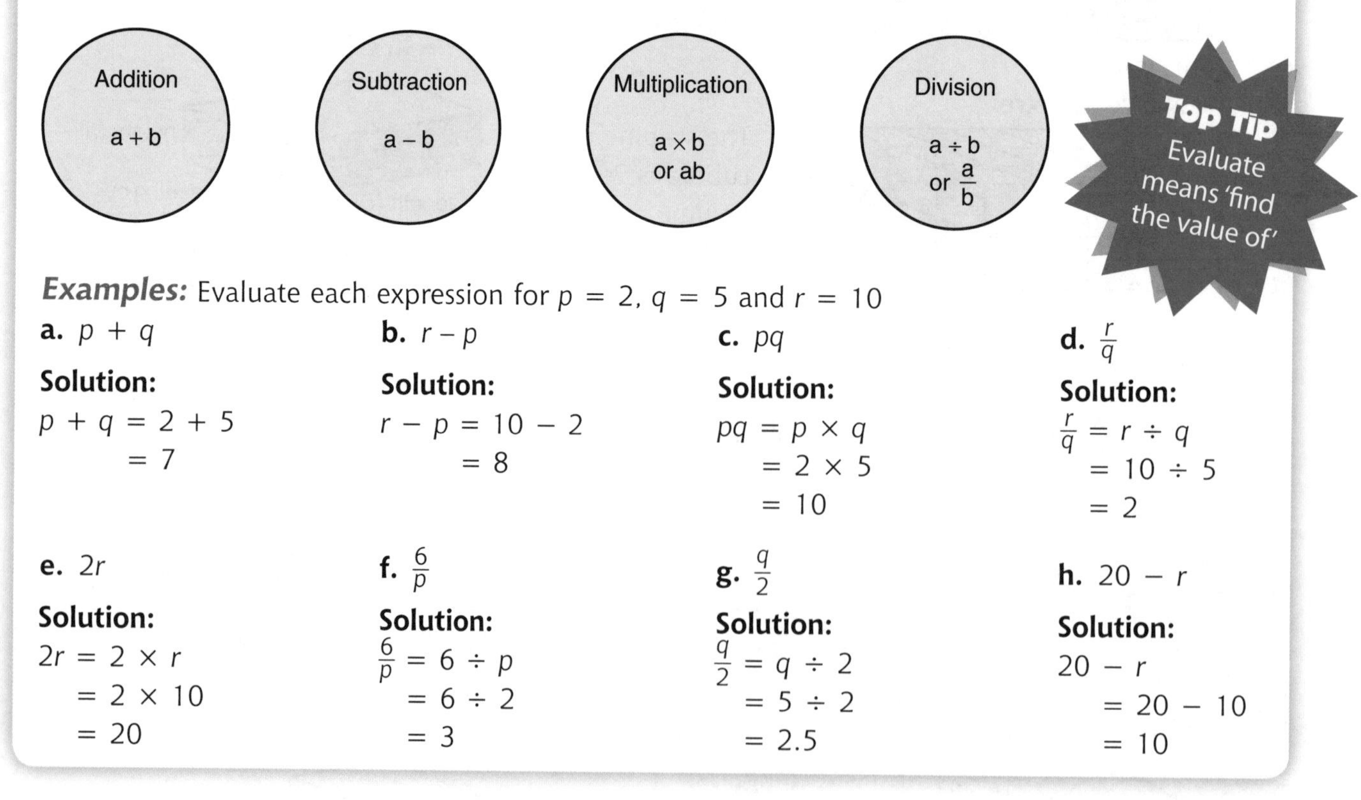

Using Letters – The Basics

Here are the four basic calculations and how they are written when you use letters for numbers:

Examples: Evaluate each expression for $p = 2$, $q = 5$ and $r = 10$

a. $p + q$

Solution:

$p + q = 2 + 5$
$= 7$

b. $r - p$

Solution:

$r - p = 10 - 2$
$= 8$

c. pq

Solution:

$pq = p \times q$
$= 2 \times 5$
$= 10$

d. $\frac{r}{q}$

Solution:

$\frac{r}{q} = r \div q$
$= 10 \div 5$
$= 2$

e. $2r$

Solution:

$2r = 2 \times r$
$= 2 \times 10$
$= 20$

f. $\frac{6}{p}$

Solution:

$\frac{6}{p} = 6 \div p$
$= 6 \div 2$
$= 3$

g. $\frac{q}{2}$

Solution:

$\frac{q}{2} = q \div 2$
$= 5 \div 2$
$= 2.5$

h. $20 - r$

Solution:

$20 - r$
$= 20 - 10$
$= 10$

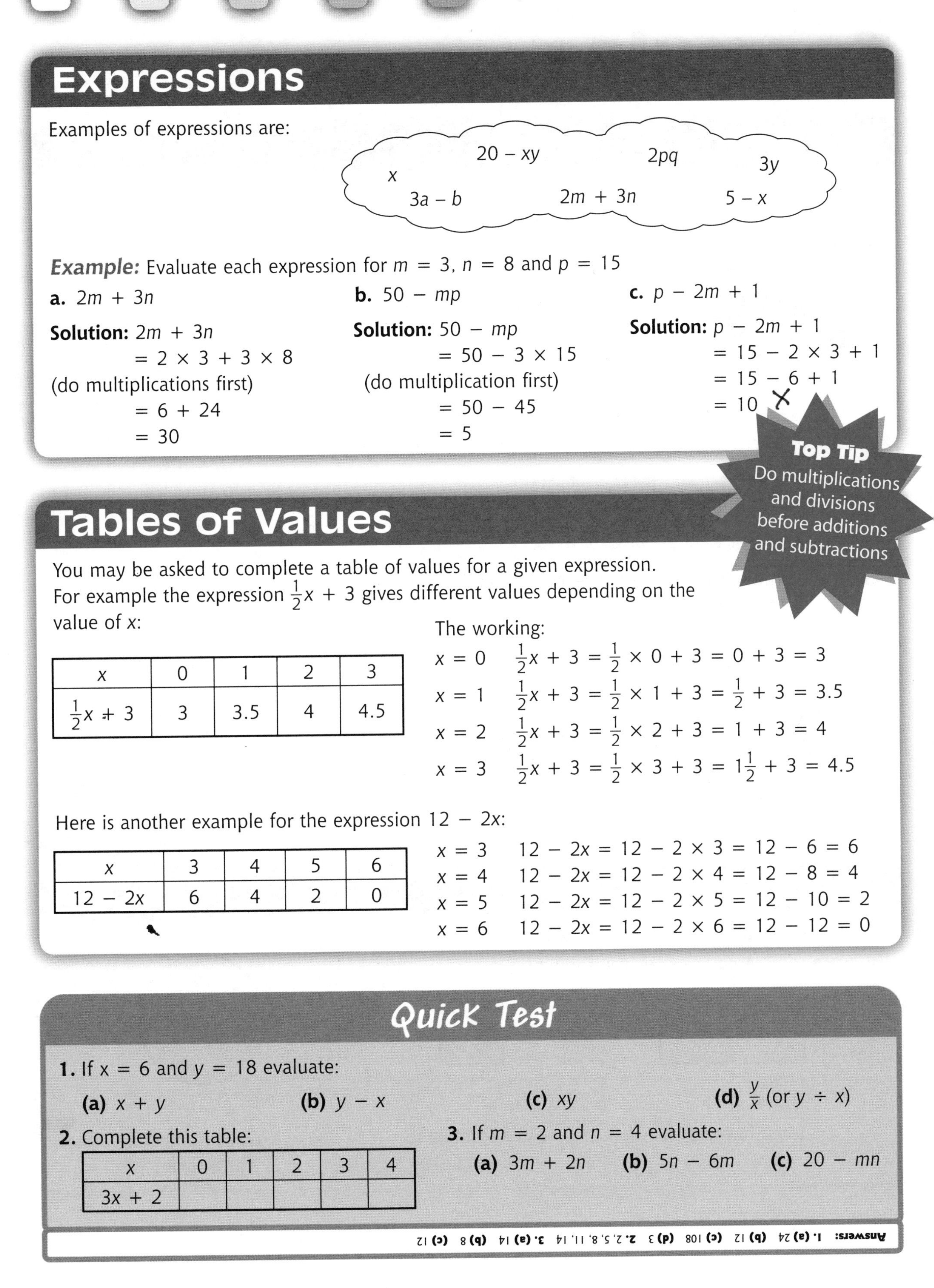

Expressions

Examples of expressions are:

x $\quad 20 - xy \quad 2pq \quad 3y \quad 3a - b \quad 2m + 3n \quad 5 - x$

Example: Evaluate each expression for $m = 3$, $n = 8$ and $p = 15$

a. $2m + 3n$

Solution: $2m + 3n$
$= 2 \times 3 + 3 \times 8$
(do multiplications first)
$= 6 + 24$
$= 30$

b. $50 - mp$

Solution: $50 - mp$
$= 50 - 3 \times 15$
(do multiplication first)
$= 50 - 45$
$= 5$

c. $p - 2m + 1$

Solution: $p - 2m + 1$
$= 15 - 2 \times 3 + 1$
$= 15 - 6 + 1$
$= 10$

Top Tip
Do multiplications and divisions before additions and subtractions

Tables of Values

You may be asked to complete a table of values for a given expression. For example the expression $\frac{1}{2}x + 3$ gives different values depending on the value of x:

x	0	1	2	3
$\frac{1}{2}x + 3$	3	3.5	4	4.5

The working:

$x = 0 \quad \frac{1}{2}x + 3 = \frac{1}{2} \times 0 + 3 = 0 + 3 = 3$

$x = 1 \quad \frac{1}{2}x + 3 = \frac{1}{2} \times 1 + 3 = \frac{1}{2} + 3 = 3.5$

$x = 2 \quad \frac{1}{2}x + 3 = \frac{1}{2} \times 2 + 3 = 1 + 3 = 4$

$x = 3 \quad \frac{1}{2}x + 3 = \frac{1}{2} \times 3 + 3 = 1\frac{1}{2} + 3 = 4.5$

Here is another example for the expression $12 - 2x$:

x	3	4	5	6
$12 - 2x$	6	4	2	0

$x = 3 \quad 12 - 2x = 12 - 2 \times 3 = 12 - 6 = 6$

$x = 4 \quad 12 - 2x = 12 - 2 \times 4 = 12 - 8 = 4$

$x = 5 \quad 12 - 2x = 12 - 2 \times 5 = 12 - 10 = 2$

$x = 6 \quad 12 - 2x = 12 - 2 \times 6 = 12 - 12 = 0$

Quick Test

1. If $x = 6$ and $y = 18$ evaluate:

(a) $x + y$ **(b)** $y - x$ **(c)** xy **(d)** $\frac{y}{x}$ (or $y \div x$)

2. Complete this table:

x	0	1	2	3	4
$3x + 2$					

3. If $m = 2$ and $n = 4$ evaluate:

(a) $3m + 2n$ **(b)** $5n - 6m$ **(c)** $20 - mn$

Answers: **1. (a)** 24 **(b)** 12 **(c)** 108 **(d)** 3 **2.** 2, 5, 8, 11, 14 **3. (a)** 14 **(b)** 8 **(c)** 12

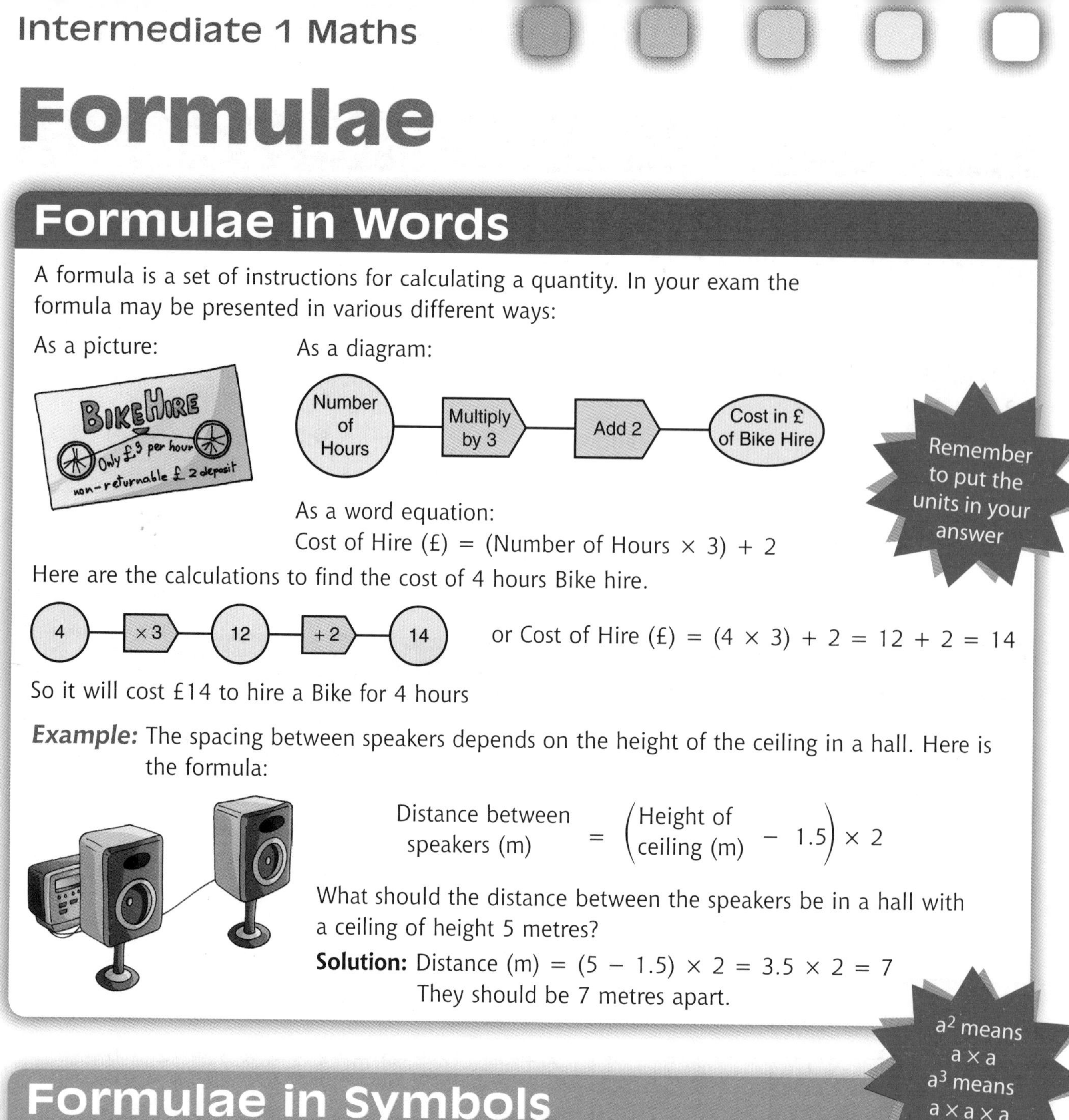

Formulae

Formulae in Words

A formula is a set of instructions for calculating a quantity. In your exam the formula may be presented in various different ways:

As a picture:

As a diagram:

As a word equation:

Cost of Hire (£) = (Number of Hours × 3) + 2

Remember to put the units in your answer

Here are the calculations to find the cost of 4 hours Bike hire.

or Cost of Hire (£) = (4 × 3) + 2 = 12 + 2 = 14

So it will cost £14 to hire a Bike for 4 hours

Example: The spacing between speakers depends on the height of the ceiling in a hall. Here is the formula:

Distance between speakers (m) = (Height of ceiling (m) − 1.5) × 2

What should the distance between the speakers be in a hall with a ceiling of height 5 metres?

Solution: Distance (m) = (5 − 1.5) × 2 = 3.5 × 2 = 7

They should be 7 metres apart.

a^2 means $a \times a$

a^3 means $a \times a \times a$

Formulae in Symbols

A formula may also be given in symbols.

Here are some Area formulae you should be familiar with:

b

ℓ

$A = \ell b$

(Area formula for a rectangle)

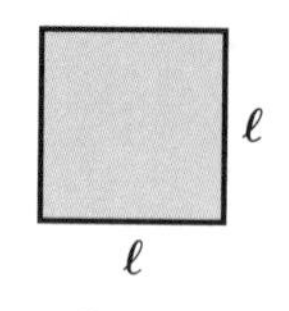

$A = \ell^2 = \ell \times \ell$

(Area formula for a square)

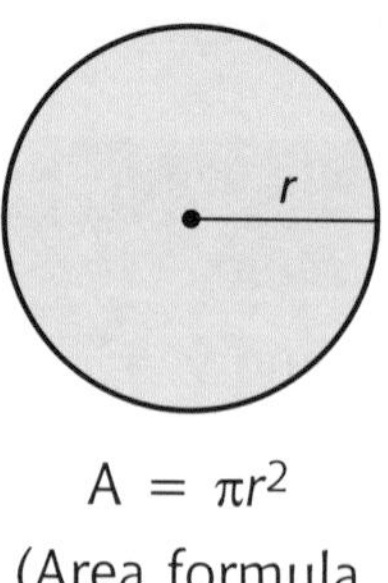

$A = \pi r^2$

(Area formula for a circle)

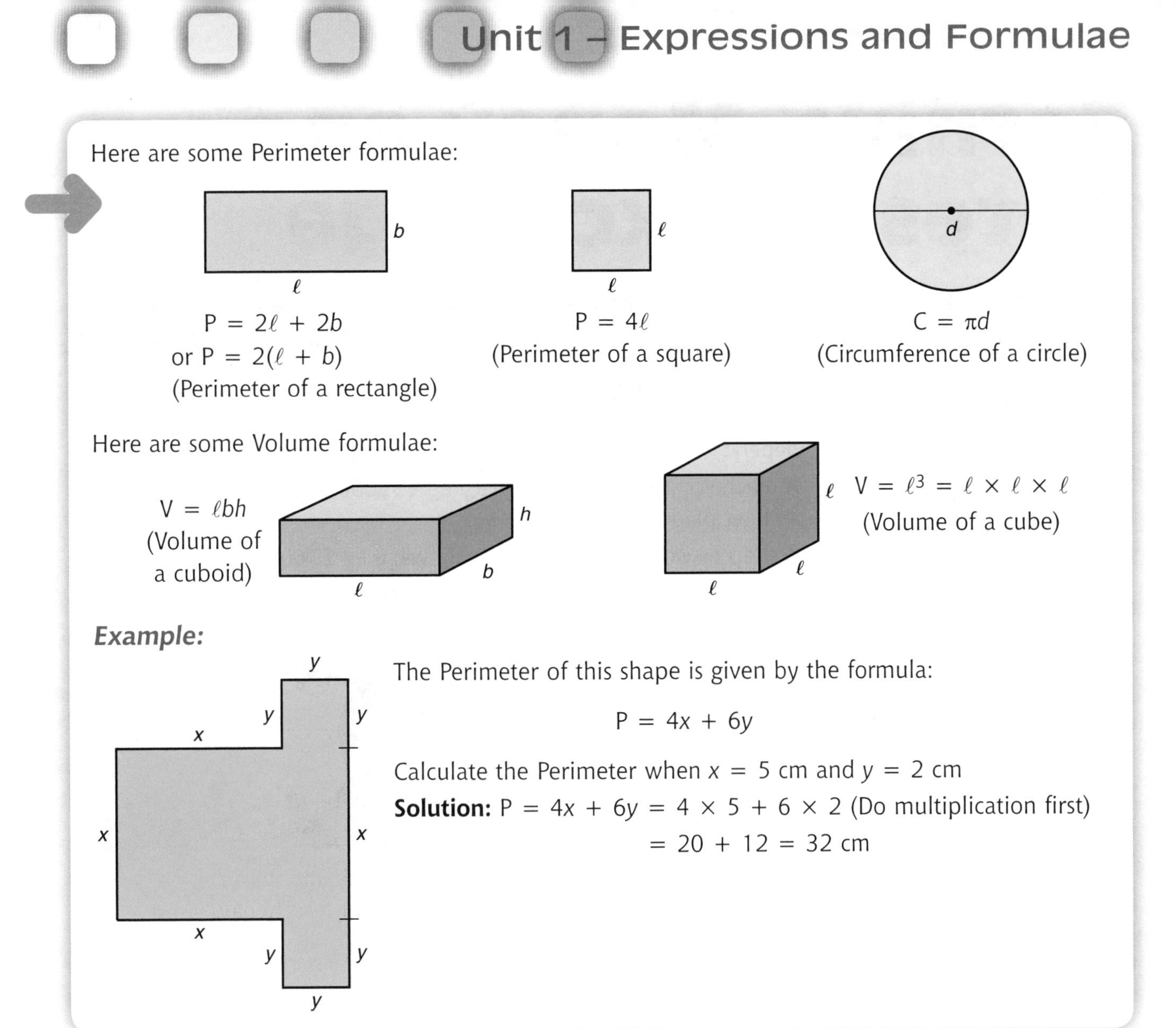

Here are some Perimeter formulae:

$P = 2\ell + 2b$
or $P = 2(\ell + b)$
(Perimeter of a rectangle)

$P = 4\ell$
(Perimeter of a square)

$C = \pi d$
(Circumference of a circle)

Here are some Volume formulae:

$V = \ell bh$
(Volume of a cuboid)

$V = \ell^3 = \ell \times \ell \times \ell$
(Volume of a cube)

Example:

The Perimeter of this shape is given by the formula:

$$P = 4x + 6y$$

Calculate the Perimeter when $x = 5$ cm and $y = 2$ cm

Solution: $P = 4x + 6y = 4 \times 5 + 6 \times 2$ (Do multiplication first)
$= 20 + 12 = 32$ cm

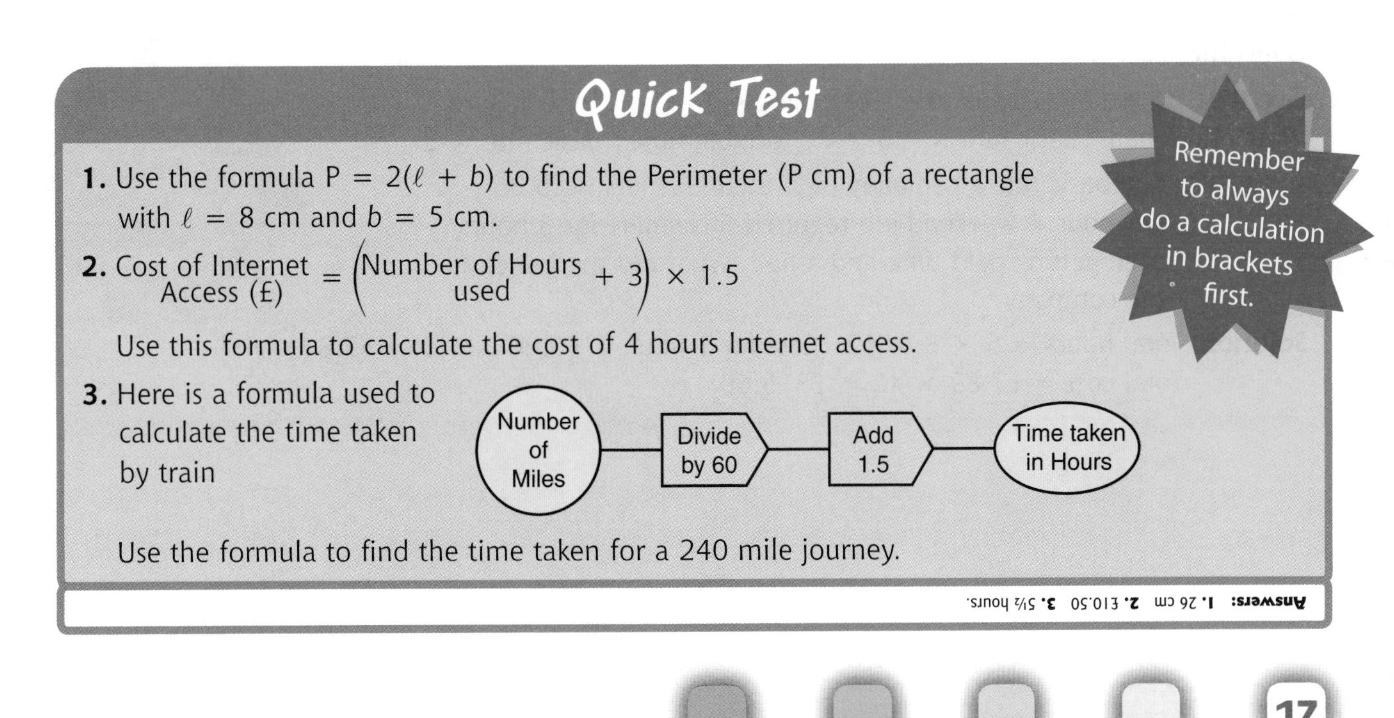

Quick Test

1. Use the formula $P = 2(\ell + b)$ to find the Perimeter (P cm) of a rectangle with $\ell = 8$ cm and $b = 5$ cm.
2. Cost of Internet Access (£) = (Number of Hours used + 3) × 1.5

   Use this formula to calculate the cost of 4 hours Internet access.
3. Here is a formula used to calculate the time taken by train

   Use the formula to find the time taken for a 240 mile journey.

Answers: **1.** 26 cm **2.** £10.50 **3.** 5½ hours.

Earning Money and Rates of Exchange

Ways of Earning Money

People are employed to do work. They get paid for the work they do. How their pay is calculated depends on the kind of work. For instance:

A teacher earns an annual salary eg £35000 and will get paid once a month.

A shop assistant will earn a weekly wage eg £360.

A seasonal farm worker may get paid 'piecework'. This means you are paid for the amount of work you do eg £10 for a box of fruit picked.

A gardener will be paid by the hour eg £9 per hour.

Note that for various reasons you will not get all of your earned money – for instance tax will be taken off your pay.

Top Tip
1 year : 12 months
1 year : 52 weeks
Annual is for a year

Here are some examples of calculating pay:

Luke has an annual salary of £27000. What does he earn each month?

Solution: There are 12 months in a year

Monthly pay = $\frac{27000}{12}$ = £2250

Zoe earns £520 per week and works for 40 hours. What is her hourly pay?

Solution: $\frac{520}{40} = 13$

She earns £13 per hour.

Ways of Earning Extra Money – Overtime

Basic Rate: normal pay for 1 hour's work

Overtime: extra hours worked

Overtime is paid at a higher rate. For example:

'time and a half': Basic rate × 1.5 'double time': Basic rate × 2

Top Tip
Time and a half– multiply by 1.5
Double Time – multiply by 2

Example: The basic rate a company pays their cleaners is £5.26 per hour. A weekend job required 5 cleaners for 8 hours each getting paid time-and-a-half. What did the job cost the company?

Solution: Total hours = 5 × 8 = 40 Cost for 1 hour = £5.26 × 1.5 = £7.89

Total cost = £7.89 × 40 = £315.60

Ways of Earning Extra Money – Commission

If your job is to sell goods you may be paid extra money called Commission. This will be a percentage of the value of the goods you sell.

Example: Peter sells computers. He earns £300 per month plus 3% commission on his monthly sales. How much does he get paid for July if he sold £52000 worth of computers that month?

Solution: Commission (Extra pay) = 3% of £52000 = £1560
Monthly pay = £300 (does not depend on sales)
Total pay for July = £300 + £1560 = £1860

Top Tip
Commission and Overtime are additions to your regular pay.

Rates of Exchange

Currency is a country's money: Scotland – Pound £
United States – Dollar $
Most European Countries – Euro €

There are well over a hundred different currencies in use in the world. The Rate of Exchange tells you how much of a currency you can buy for £1. These Rates change day-by-day.

Top Tip
A Rate of Exchange tells you how much £1 is worth.

Example:

This camera was priced: $230 in a US site and €165 on a French site.
Rates of Exchange at the time were £1 = $1.93 = €1.25.
On which site was it cheapest?

Solution: Change each price to £

US: $\frac{230}{1.93} = £119.17$

France: $\frac{165}{1.25} = £132.00$

It is cheapest on the US site.

Example: Tim travels to Boston in the US at a time when the exchange rate is £1=$1.36. He takes with him £500. How many dollars does he receive for his money?

Solution: Each pounds is worth $1.36 so
£500 = $1.36 × 500
= $1.36 × 500
= $680

Note: Here is a useful rule when changing currencies:

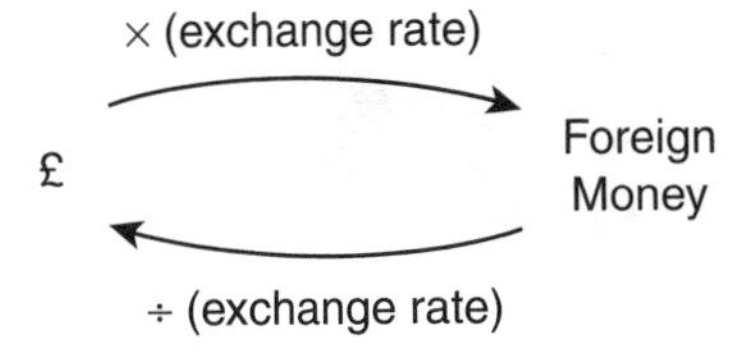

Quick Test

1. A night-shift worker gets paid 'time-and-a-half'. If the basic rate is £5.40 per hour and she worked 6 hours on night shift, what was her pay?
2. Tim earns £600 per month plus 2% commission on sales. If his sales were £40000 one month, what did he earn that month?

Answers: 1. £48.60 2. £1400

Insurance and Higher Purchase

What is Insurance?

Accidents happen –

your car crashes

your house goes on fire

Accidents cost you money to fix. An insurance company will pay these costs if you pay them monthly or annually for insurance cover. The payments you make are called PREMIUMS.

Buildings and Contents Insurance

Top Tip
Premium quotes are often given per £1000 of cover.

To protect yourself against accidents to your house– fires, floods, gales etc. – you buy 'building insurance'. The premium is paid annually and is calculated for each £1000 your house is worth. In a similar way the contents of your house can be insured against theft or damage, etc.

Example: Peter insures his house, worth £140000, with Star Insurance Ltd. They charge £2.50 per £1000. Calculate his annual premium.

Solution: £140000 = 140 × £1000 (140 lots of £1000)

Annual Premium = 140 × £2.50 (140 lots of £2.50 since each £1000 costs £2.50)
= £350

Car Insurance

Top Tip
Calculating Percentages: Find 1% by dividing by 100 Then multiply to find the required percentage.

It is illegal to drive a car that is not insured. The cost of car insurance depends on the type of car, your age, where you live etc.

You get a reduction on your premiums, called 'no-claims discount', if you have not claimed money from the insurance company for a few years.

Example: Yvonne's annual premium is £450 but she has not claimed for 3 years. What is her premium with discount?

No. of Years (no claims)	Discount on Premiums
1	25%
2	30%
3	35%
4	40%

Solution: Discount = 35% of £450
= 450 ÷ 100 × 35
= £157.50

Annual Premium = £450 − £157.50
= £292.50

Holiday Insurance

If your holiday is cancelled, or you lose your luggage or you fall ill your holiday insurance will cover the costs.

Example:

Mr & Mrs Curry take their 8 year old son on a 2 week holiday and use the Gold insurance policy. Calculate their insurance premium.

SUN HOLIDAY INSURANCE		
Length of Holiday	**Silver Adult/Child**	**Gold Adult/Child**
up to 9 days	£25/£15	£33/£18
10 days to 17 days	£30/£20	£39/£24
18 days to 31 days	£35/£25	£45/£30

Solution: 2 weeks is 14 days (middle row of table)
so the costs are 1 Adult: £39   1 Child: £24

Total premium for 2 Adults and 1 Child = 2 × £39 + £24
= £78 + £24
= £102

Hire Purchase

Hire Purchase or HP is a method of paying for expensive goods.

For example: You pay a DEPOSIT eg 10% of the cash price: 10% of £500 = £50

You then pay regular INSTALMENTS eg £65 each month for a year: 12 × £65 = £780

The total HP Price = £50 (Deposit) + £780 (Instalments) = £830

TV price: £500 cash

Note: This HP Price is £330 more than the cash price £500.

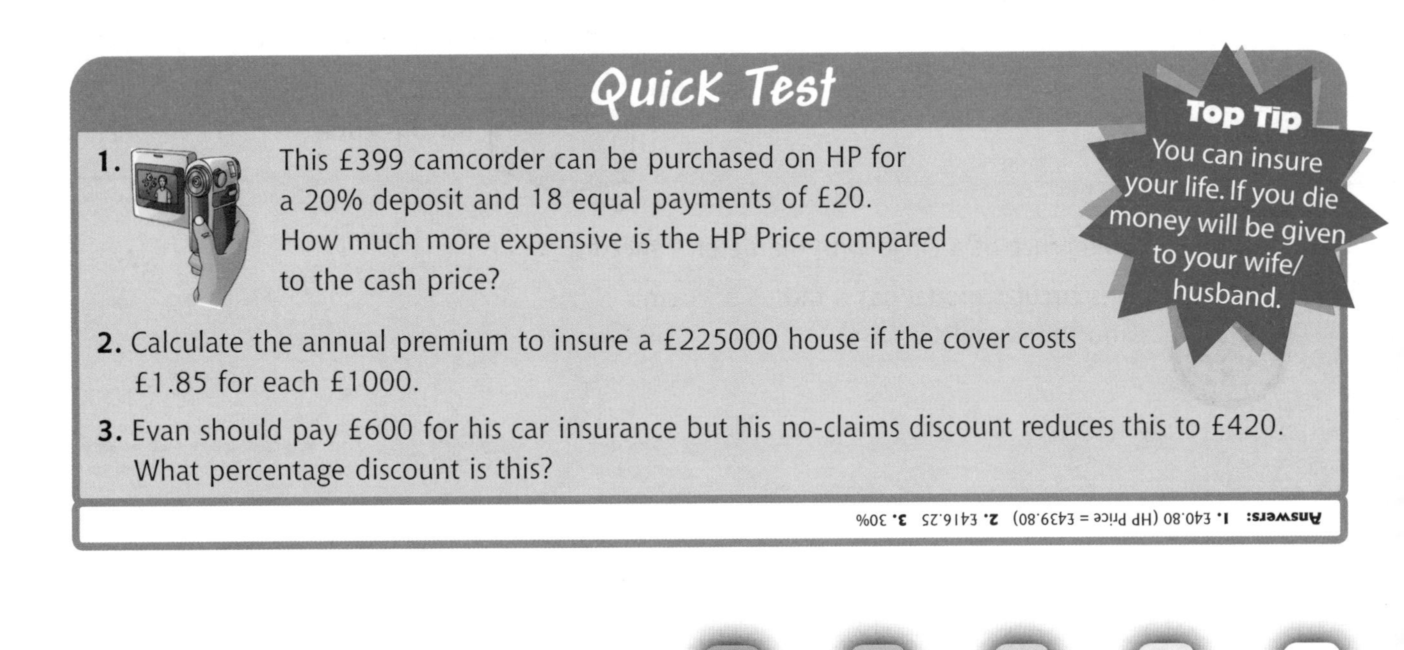

Quick Test

1. This £399 camcorder can be purchased on HP for a 20% deposit and 18 equal payments of £20. How much more expensive is the HP Price compared to the cash price?
2. Calculate the annual premium to insure a £225000 house if the cover costs £1.85 for each £1000.
3. Evan should pay £600 for his car insurance but his no-claims discount reduces this to £420. What percentage discount is this?

Top Tip
You can insure your life. If you die money will be given to your wife/husband.

Answers: **1.** £40.80 (HP Price = £439.80) **2.** £416.25 **3.** 30%

Practice Unit 1 Test

Formulae

You will be given these formulae in your Unit 1 Assessments:

Circumference of a circle: $C = \pi d$

Area of a circle: $A = \pi r^2$

Outcome 1

1. Mr & Mrs Ross paid £120.50 for a flight to London. They had to pay a 12% surcharge for their flight due to fuel cost rises. Calculate the surcharge. (2)

2. Hot Chilli Chocs cost £2.50 for 100 g. Calculate the VAT (at 15%) paid for 100 g. Give your answer to the nearest 1 p. (3)

3. **(a)** A car travels 125 kilometres in one hour. How far will it travel in 3 hours? Round your answer to the nearest ten km. (3)

   (b) A honey bee is observed doing a waggle dance. If it does 8 circuits in 15 seconds how many circuits per minute is this? (2)

Outcome 2

4.

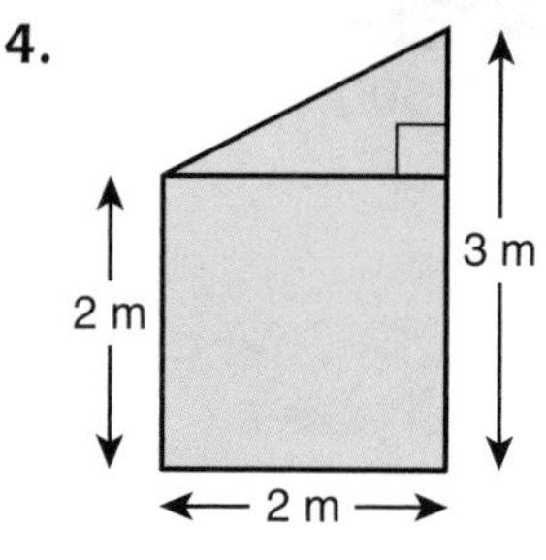

The diagram shows the cross section of a lean-to shed which consists of a square and a right-angled triangle.

Find the area of the cross-section. (5)

5. **(a)**

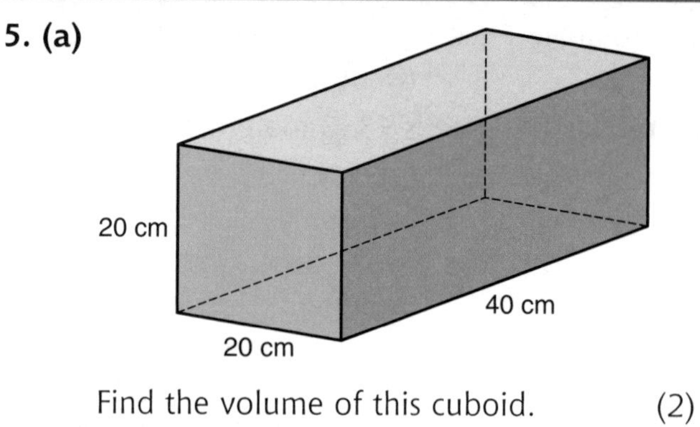

Find the volume of this cuboid. (2)

(b)

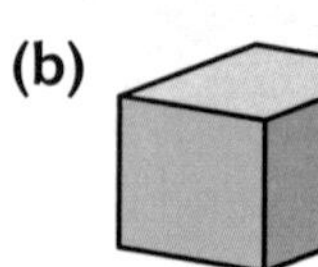

Find the volume of this cube. (2)

6. Find the circumference of a circle with diameter 9 metres. (2)

7. This circular medal has a radius of 3 cm. Find the medal's area.

Outcome 3

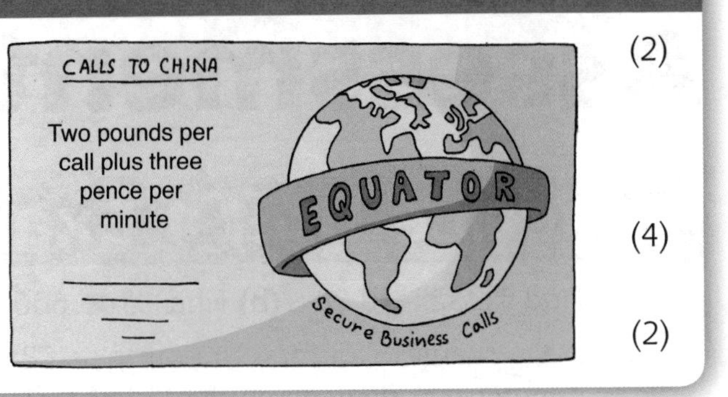

8. Evaluate $4x - 3y$ when $x = 8$ and $y = 4$. (2)

9. A businessman used 'Equator' to make 8 calls to China which lasted a total of 390 minutes. How much did the calls cost him altogether? (4)

10. For the formula $D = ST$ evaluate D when $S = 30$ and $T = 4$. (2)

Outcome 4

11. 

Lewis is a cleaner. He is paid a basic rate of £6.32 per hour. Sometimes he works night shift and is paid at time and a half. One week he worked 23 hours on night shift. How much did he get paid? (4)

12. Find the cost of buying these binoculars on H.P.

BINOCULARS
on H.P.
A fabulous deal!!

DEPOSIT ONLY £25
18 easy instalments of £5.50 (3)

13.

EXCHANGE RATE TABLE

£1 buys:	
China	13.7 yuan
USA	2.13 dollars
France	1.26 euro

Mr Johnstone is travelling to China to take up a teaching post. He has £630 that he wants to exchange for yuan.

How many yuans will he get for his money?

Preparation for Assessment

Practice Unit 1 Exam Questions

1. **(a)** Find 1.932 − 0.28 **(b)** Find 13 × 600 **(c)** Find 8.46 ÷ 3

2. A garage charges £25 for a call-out plus £8 for every 20 minutes that they are on the call-out. How much do they charge for a 3 hour call-out?

3. Megan invests £2860 in a savings account at 6% per annum. Calculate the interest she receives after 4 months.

4. The table shows the annual insurance premium for car insurance charged by Road Safe Ltd.

ANNUAL PREMIUMS

	Age in Years			
	18–20	21–25	26–35	over 35
Area A	£826	£632	£440	£380
Area B	£1050	£750	£582	£420
Area C	£1280	£895	£630	£510

(a) Megan uses Road Safe Ltd. She is 28 years old and lives in Area B. What is her annual premium?

(b)

Number of years without a claim	1	2	3	4
Discount	25%	35%	40%	45%

Her 24 year old friend Abi lives in Area A and has not made a claim for 3 years and is entitled to a discount on her premium as shown in the table.

How much does it cost Abi to insure her car?

5. The weight of books is proportional to their volume.

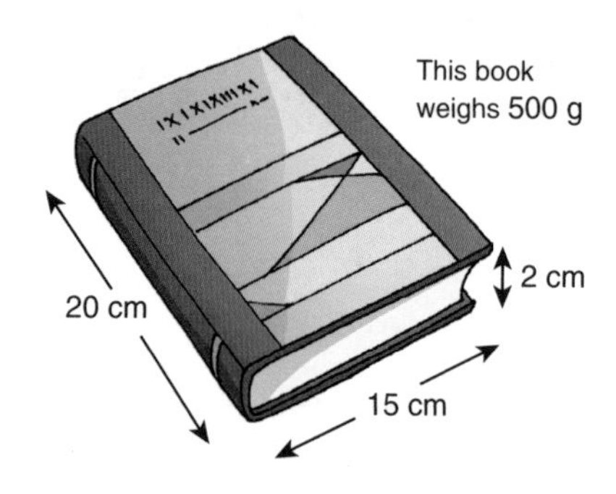

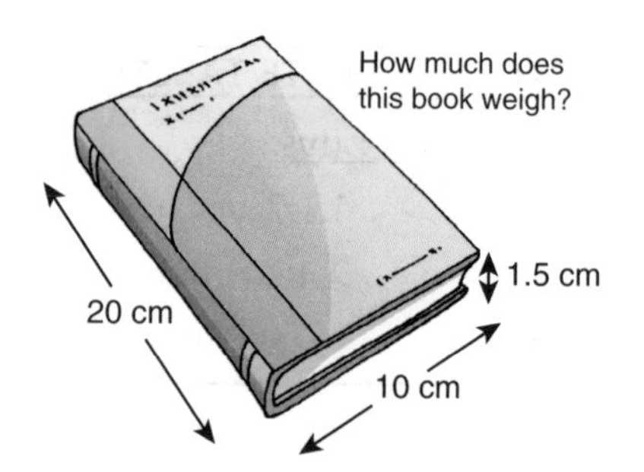

6.

CONCERT TICKET PRICES		
	Adult	Child
Front Area	€35	€20
Side Area	€25	€15
Rear Area	€20	€10

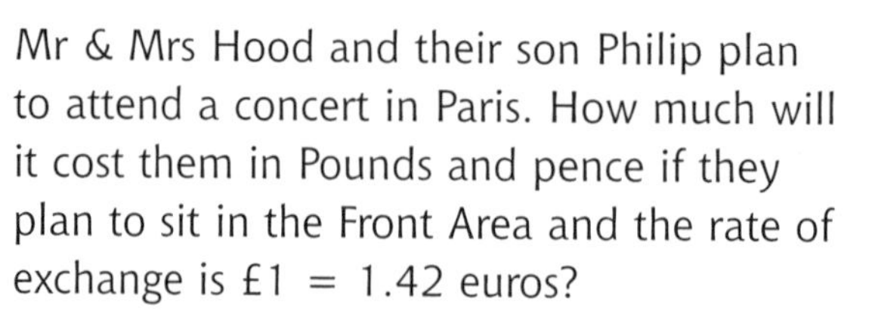

Mr & Mrs Hood and their son Philip plan to attend a concert in Paris. How much will it cost them in Pounds and pence if they plan to sit in the Front Area and the rate of exchange is £1 = 1.42 euros?

7. A T.V. that originally cost £320 is now being sold in a sale at adiscounted price of £280. Express the discount as a percentage of the original price.

8. Find 80% of £270.

9. 

his wants to insure his collection of Star War Figures which is worth £6000. The Insurance Company charges an annual premium of £8.42 for each £1000 insured. Work out his annual premium.

10. 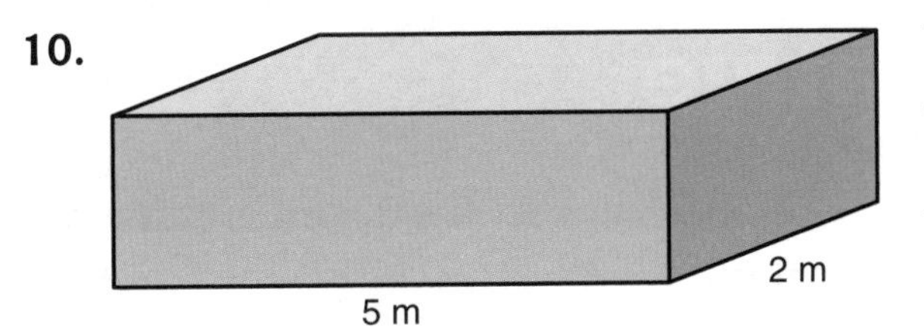

This petrol tank is shaped as a cuboid. If it holds 15 m^3 of petrol work out the height of the tank.

11. Jack has an annual salary of £28600. He is awarded a pay rise of 2.6%. Calculate his new annual salary after tax of 20% has been deducted.

12.

EXCHANGE RATES
£1 = 1.32 euro £1 = 220 yen

Namiko travelled from Tokyo to Edinburgh where she changed 264000 yen into pounds. She spent £350 and changed the rest of her money into euros before she travelled to France.

How many euros did she get?

Working with Integers

What are Integers?

Integers are positive or negative whole numbers including zero.
Here they are on a number line:

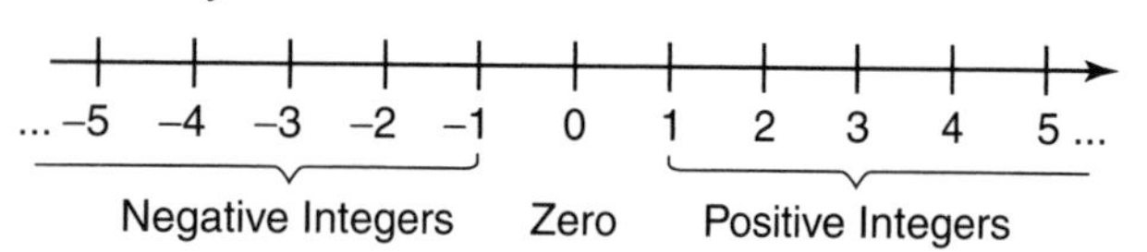

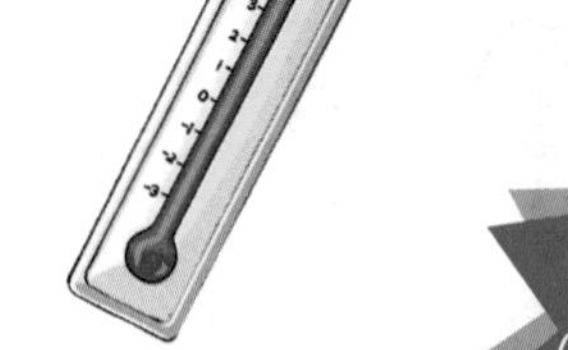

Top Tip
Plotting Points:
(left/right, up/down)

Plotting points – negative coordinates

To describe the position of a Point you need two coordinates:

The x-coordinate tells you left/right

The y-coordinate tells you up/down

P (x, y)

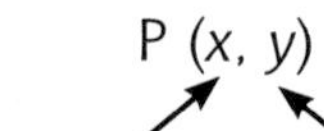

x-coordinate (1st number)

y-coordinate (2nd number)

Here are four points plotted on a coordinate diagram:

B (–2, 1) A (2, 1) O C (–2, –1) D (2, –1) x y

⇐ y-coordinates (2nd numbers) are positive (up from 0)

⇐ y-coordinates (2nd numbers) are negative (down from 0)

⇑ x-coordinates (1st numbers) are negative (left from 0)

⇑ x-coordinates (1st numbers) are positive (right from 0)

For example starting from 0 you will find:

A (2, 1) two squares right and one square up from 0.

B (–2, 1) two squares left and one square up from 0.

C (–2, –1) two squares left and one square down from 0.

D (2, –1) two squares right and one square down from 0.

Integers in Context

Here is a Thermometer:

The temperature is –1°C
(1 degree below zero)

Note: –2°C is below –1°C

So –2°C is colder than –1°C

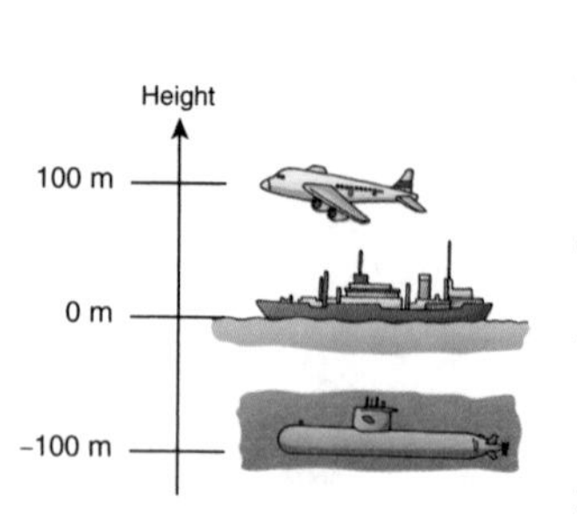

The plane is 100 m above sea-level

The ship is at sea-level

The submarine is 100 m below sea-level.

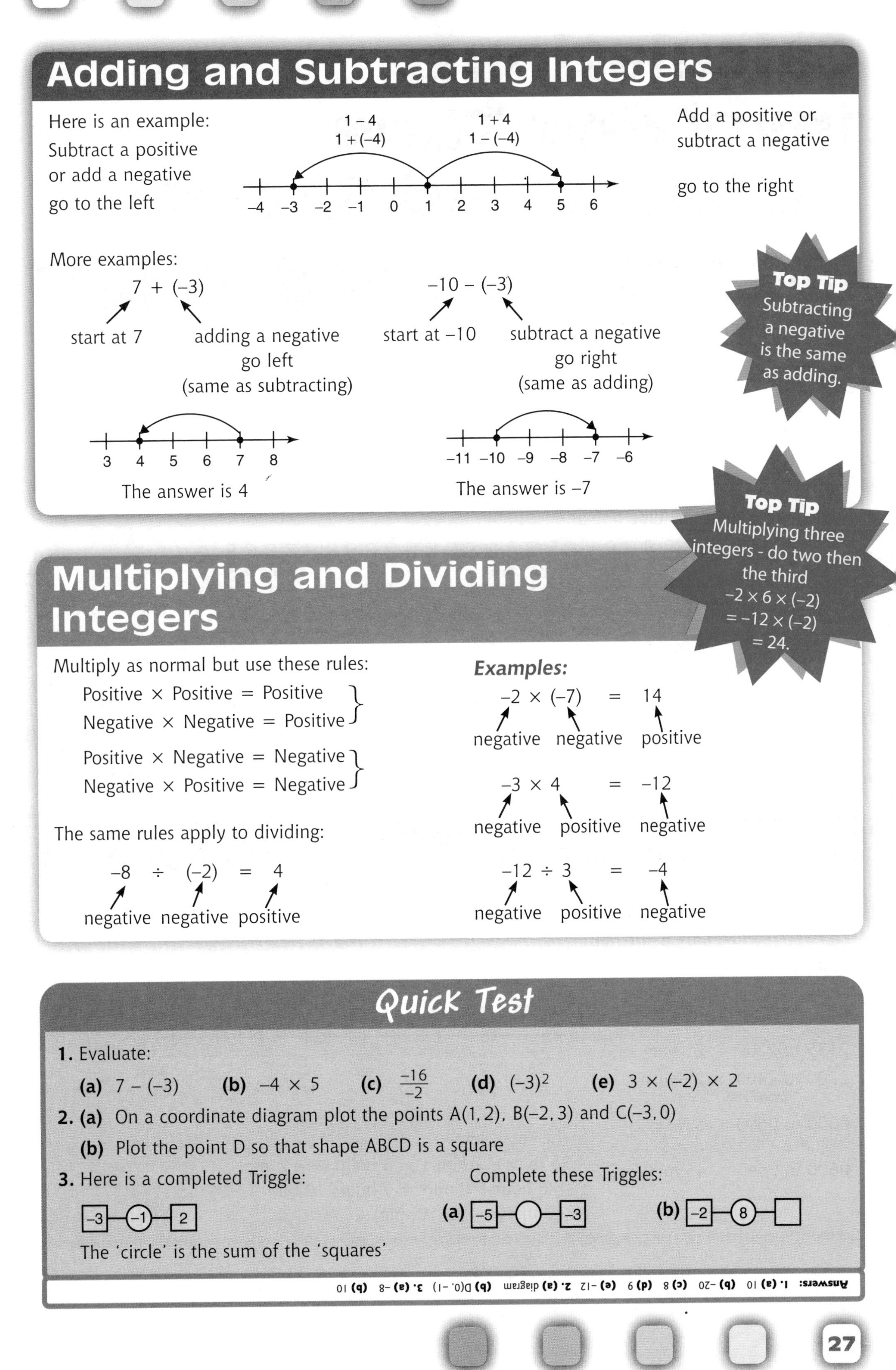

Adding and Subtracting Integers

Here is an example:

Subtract a positive or add a negative go to the left

Add a positive or subtract a negative go to the right

More examples:

$7 + (-3)$

start at 7; adding a negative go left (same as subtracting)

The answer is 4

$-10 - (-3)$

start at −10; subtract a negative go right (same as adding)

The answer is −7

Top Tip
Subtracting a negative is the same as adding.

Multiplying and Dividing Integers

Top Tip
Multiplying three integers - do two then the third
$-2 \times 6 \times (-2)$
$= -12 \times (-2)$
$= 24.$

Multiply as normal but use these rules:

Positive × Positive = Positive
Negative × Negative = Positive

Positive × Negative = Negative
Negative × Positive = Negative

The same rules apply to dividing:

$-8 \div (-2) = 4$

negative negative positive

Examples:

$-2 \times (-7) = 14$

negative negative positive

$-3 \times 4 = -12$

negative positive negative

$-12 \div 3 = -4$

negative positive negative

Quick Test

1. Evaluate:

(a) $7 - (-3)$ **(b)** -4×5 **(c)** $\frac{-16}{-2}$ **(d)** $(-3)^2$ **(e)** $3 \times (-2) \times 2$

2. (a) On a coordinate diagram plot the points A(1, 2), B(−2, 3) and C(−3, 0)

(b) Plot the point D so that shape ABCD is a square

3. Here is a completed Triggle:

[−3]—(−1)—[2]

The 'circle' is the sum of the 'squares'

Complete these Triggles:

(a) [−5]—( )—[−3] **(b)** [−2]—(8)—[ ]

Answers: 1. (a) 10 **(b)** −20 **(c)** 8 **(d)** 9 **(e)** −12 **2. (a)** diagram **(b)** D(0, −1) **3. (a)** −8 **(b)** 10

Calculations Involving Time, Distance and Speed

What Time Is It?

Here is a time line showing the times during the course of a day:

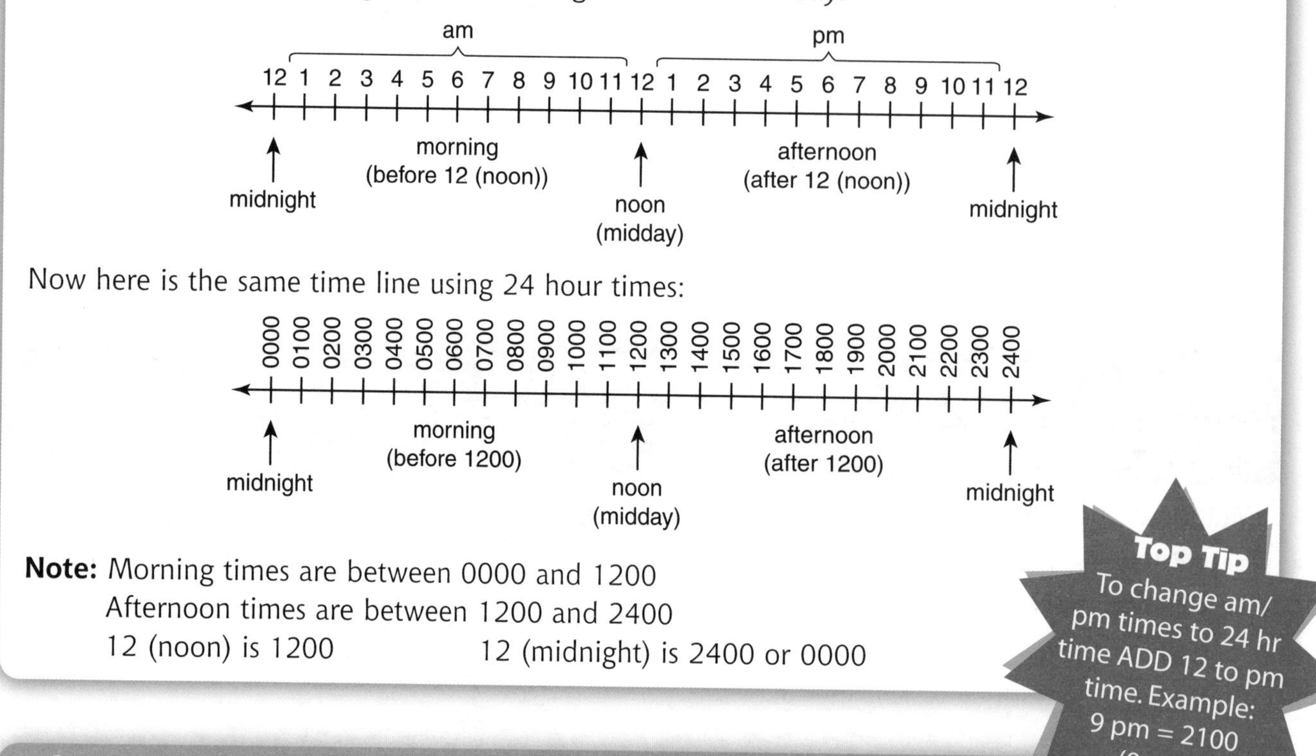

Note: Morning times are between 0000 and 1200
Afternoon times are between 1200 and 2400
12 (noon) is 1200 12 (midnight) is 2400 or 0000

Top Tip
To change am/pm times to 24 hr time ADD 12 to pm time. Example: 9 pm = 2100 (9 + 12)

Time Intervals

It is helpful to think of a time line when calculating time intervals:

Example: Rhoda works a night shift starting at 2135 and finishing at 0645. How long is her shift?

Solution:

2135 2200 2400 0600 0645
25 min 2 hrs 6 hrs 45 min

From:

2135 to 2200 : 25 min
2200 to 2400 (midnight) : 2 hours
0000 (midnight) to 0600 : 6 hours
0600 to 0645 : 45 min

The Total time is:
25 min + 2 hours + 6 hours + 45 min
= 8 hours 70 min = 9 hours 10 min
(70 min = 1 hr 10 min)

Calculating Time, Distance or Speed

To remember the three formulae use this triangle:

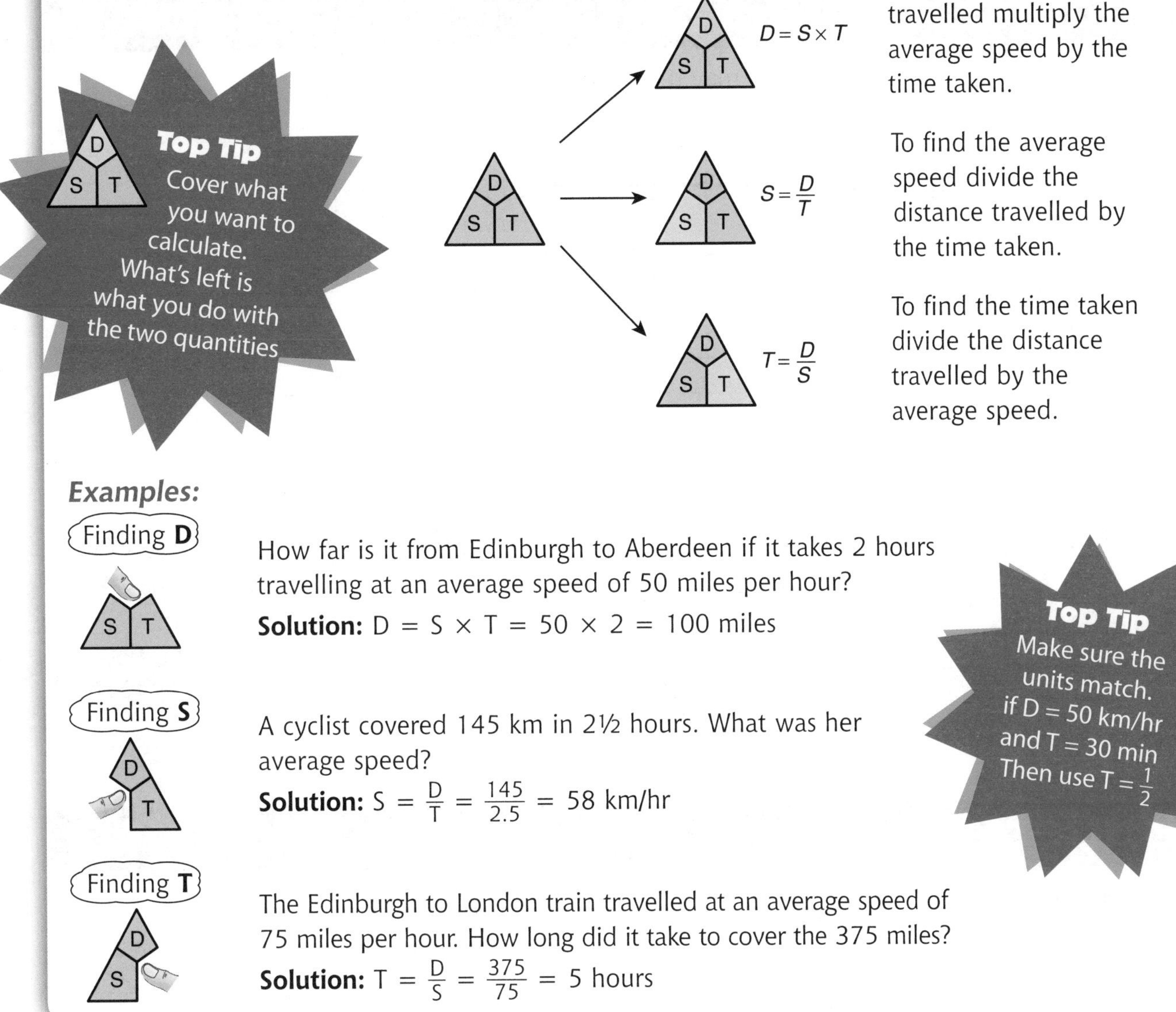

To find the distance travelled multiply the average speed by the time taken.

To find the average speed divide the distance travelled by the time taken.

To find the time taken divide the distance travelled by the average speed.

Examples:

Finding **D**

How far is it from Edinburgh to Aberdeen if it takes 2 hours travelling at an average speed of 50 miles per hour?

Solution: $D = S \times T = 50 \times 2 = 100$ miles

Finding **S**

A cyclist covered 145 km in 2½ hours. What was her average speed?

Solution: $S = \frac{D}{T} = \frac{145}{2.5} = 58$ km/hr

Finding **T**

The Edinburgh to London train travelled at an average speed of 75 miles per hour. How long did it take to cover the 375 miles?

Solution: $T = \frac{D}{S} = \frac{375}{75} = 5$ hours

Quick Test

1. The overnight train left at 2245 and arrived the next day at 0815
 (a) How long did it take?
 (b) What was the average speed over the 646 miles?

2. Angela's car journey took $1\frac{3}{4}$ hours. If her average speed was 40 miles per hour, how far did she travel?

3. Amy ran at an average speed of 6 metres per second for the 1 km race. How long did it take her? Give your answer in minutes and seconds, to the nearest second.

4. The bus took 2 hours 10 min. When will it arrive if it left at 0855?

Answers: **1. (a)** 9 hrs 30 min **(b)** 68 mph **2.** 70 miles **3.** 2 min 47 secs **4.** 1105

Distance/Time Graphs

Understanding a Distance/Time Graph

As an example, this graph shows a journey that Ryan made from his home in Glasgow to Aberdeen and back.

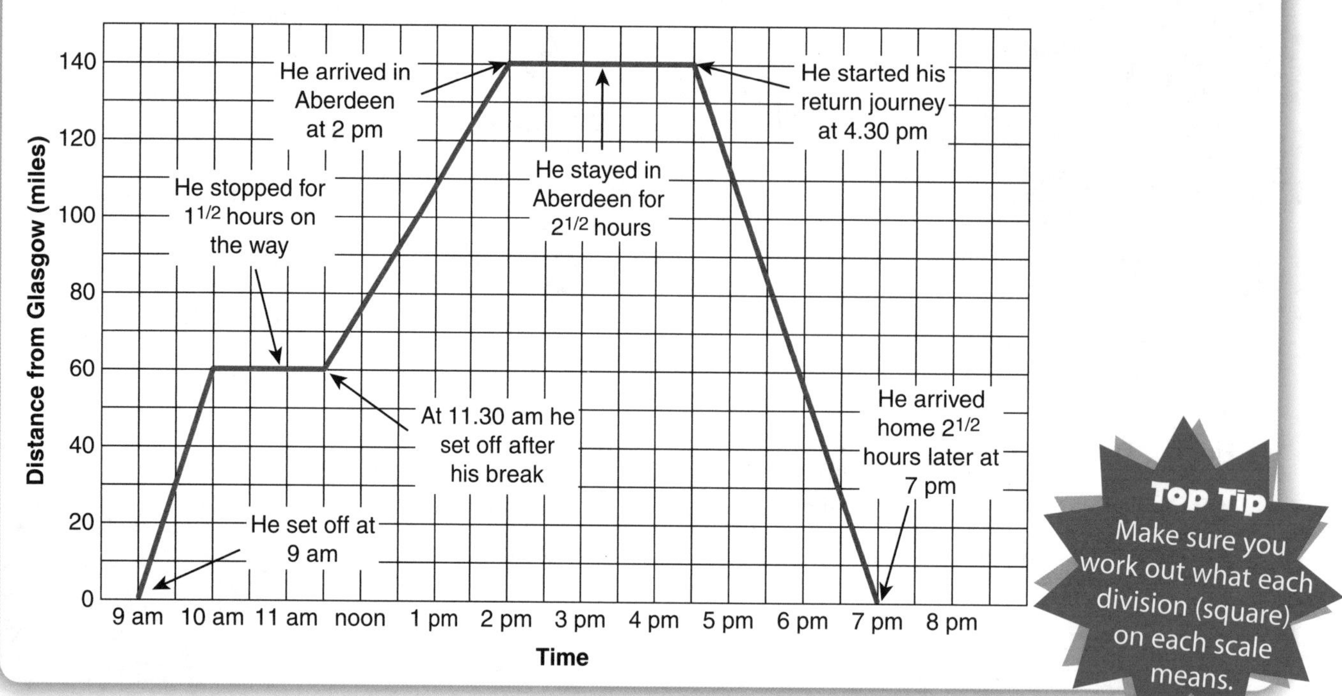

Top Tip
Make sure you work out what each division (square) on each scale means.

Understanding a Two Graph Diagram

Yvonne's train left Edinburgh for London at the same time that Leanne's train left London for Edinburgh.

The graph shows the progress of the two trains.

Distance from Edinburgh (miles)
480
400
320
240
160
80
0
Leanne's train leaves London at 9.45 am
The two trains pass each other 160 miles from Edinburgh at 1300
Yvonne arrives in London at 1600
Yvonne's train leaves Edinburgh at 9.45 am
Yvonne's train stops for 45 min
Leanne arrives in Edinburgh at 1515
0900
1000
1100
1200
1300
1400
1500
1600
Time

Scales

You need to take care with the scales used on the graphs

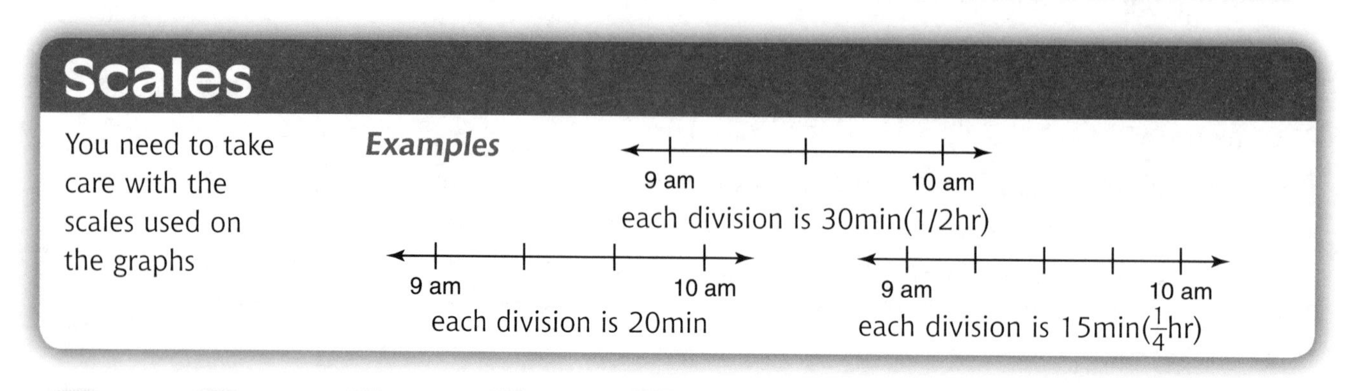

Calculating Average Speed from a Graph

Here are two examples

Tim travels from home to work

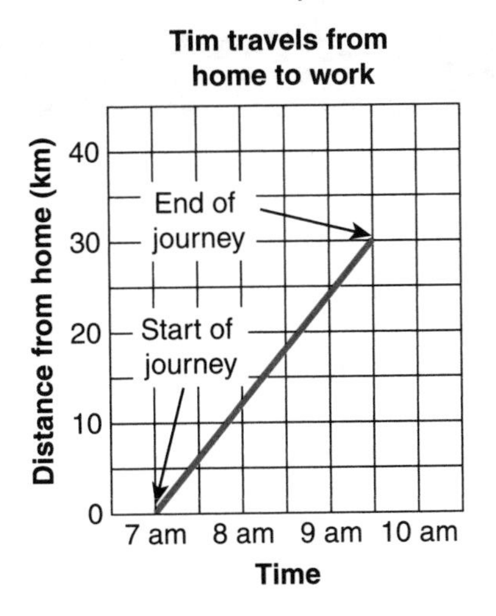

His journey started at 7 am and he arrived at 9.30 am. He took $2\frac{1}{2}$ (2.5) hours for the 30 km journey.

Now use:

$$S = \frac{D}{T} = \frac{30}{2.5} = 12 \text{ km/hr}$$

Yvonne visits her friend in a nearby town

Distance from home (miles): 0, 5, 10, 15, 20, 25

Journey to visit her friend

Her Journey home

Time: 1500, 1600, 1700, 1800

Her 15 km journey to her friend took $1\frac{1}{2}$ (1.5) hours (1500 to 1630)

So $S = \frac{D}{T} = \frac{15}{1.5} = 10$ km/hr

Her return journey took $\frac{3}{4}$ hr (0.75) (17.15 to 1800)

So $S = \frac{D}{T} = \frac{15}{0.75} = 20$ km/hr

Note: The steeper the line on a distance/time graph the faster the average speed

Top Tip

On a distance/time graph

$$\text{Av. speed} = \frac{\text{up or down (D)}}{\text{along (T)}}$$

Quick Test

Gavin travels to work each day by car. This graph shows his journey on Friday one week:

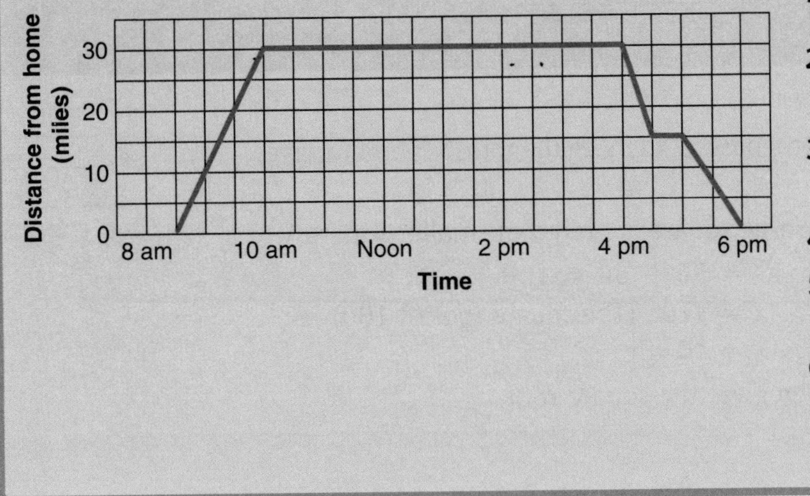

1. When did Gavin leave for work?
2. How long did he take getting to work?
3. What was his average speed going to work?
4. How long was he at work?
5. He was held up on the way home. For how long?
6. What was his average speed for the whole of his return journey?

Answers: 1. 8.30 am **2.** $1\frac{1}{2}$ hrs **3.** 20 mph **4.** 6 hrs **5.** 1/2 hr **6.** 15 mph

The Theorem of Pythagoras

Understanding Pythagoras' Theorem

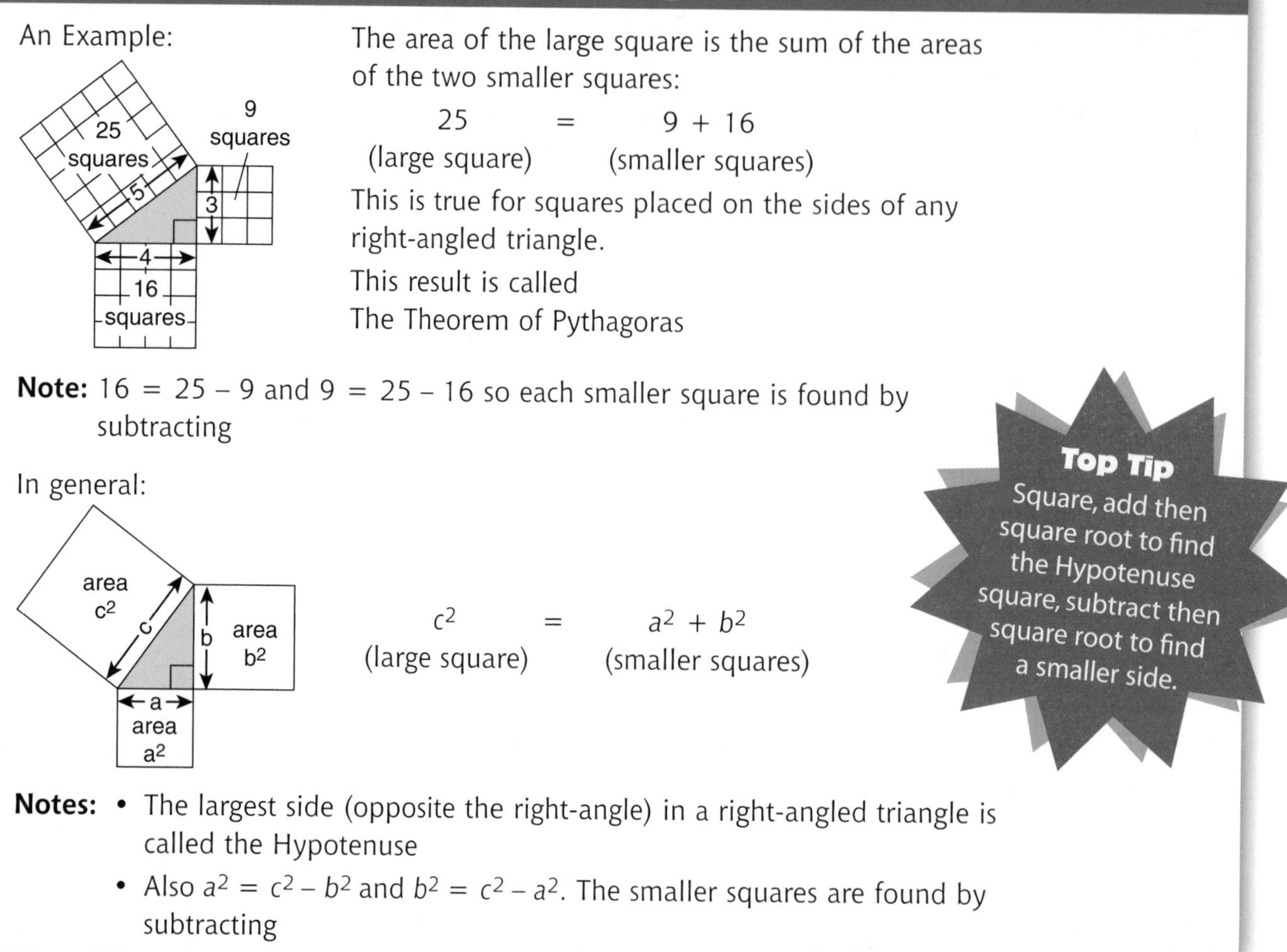

An Example:

The area of the large square is the sum of the areas of the two smaller squares:

25	=	9 + 16
(large square)		(smaller squares)

This is true for squares placed on the sides of any right-angled triangle.

This result is called
The Theorem of Pythagoras

Note: 16 = 25 – 9 and 9 = 25 – 16 so each smaller square is found by subtracting

In general:

c^2	=	$a^2 + b^2$
(large square)		(smaller squares)

Top Tip
Square, add then square root to find the Hypotenuse
square, subtract then square root to find a smaller side.

Notes:
- The largest side (opposite the right-angle) in a right-angled triangle is called the Hypotenuse
- Also $a^2 = c^2 - b^2$ and $b^2 = c^2 - a^2$. The smaller squares are found by subtracting

Calculating the Largest Side (Hypotenuse)

Example:

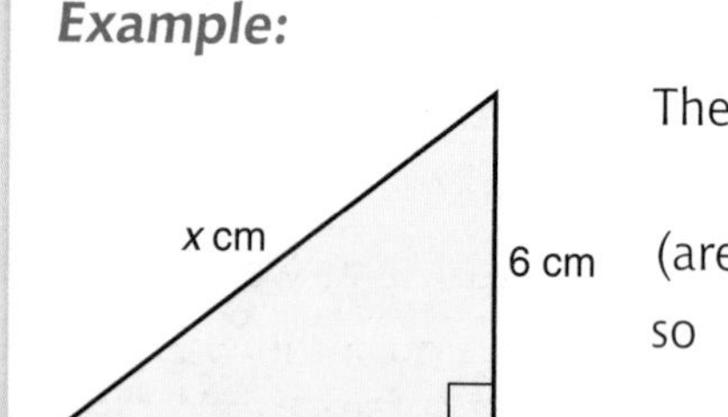

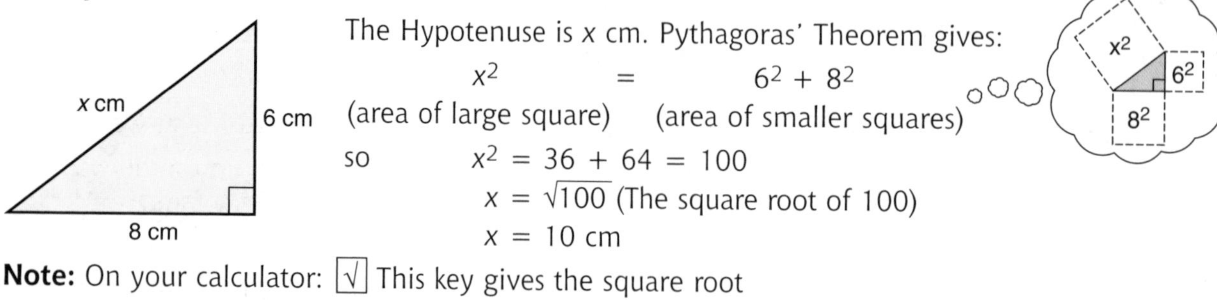

The Hypotenuse is x cm. Pythagoras' Theorem gives:

x^2	=	$6^2 + 8^2$
(area of large square)		(area of smaller squares)

so $x^2 = 36 + 64 = 100$

$x = \sqrt{100}$ (The square root of 100)

$x = 10$ cm

Note: On your calculator: $\boxed{\sqrt{}}$ This key gives the square root

Calculating a Smaller Side

Example:

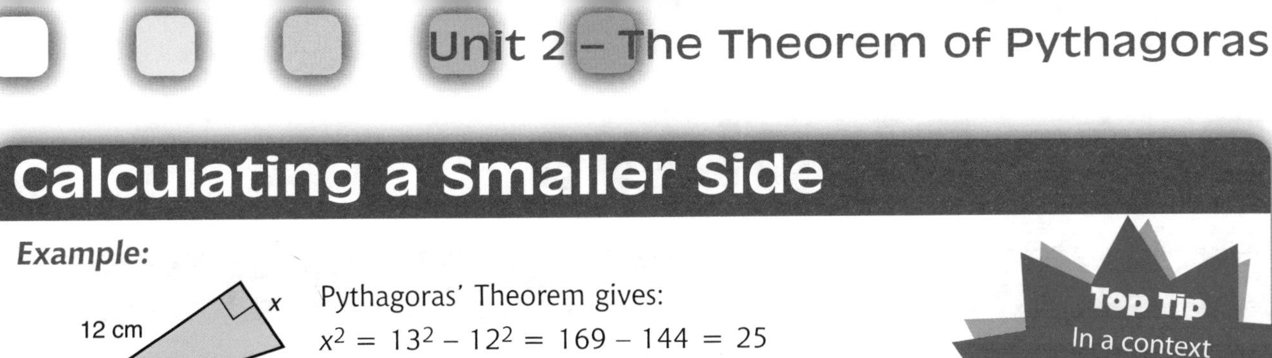

Pythagoras' Theorem gives:

$x^2 = 13^2 - 12^2 = 169 - 144 = 25$

$x = \sqrt{25}$ (The square root of 25)

$x = 5$ cm

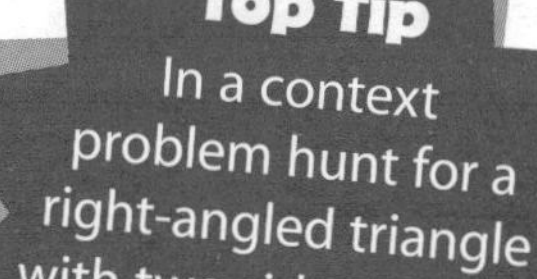

Top Tip

In a context problem hunt for a right-angled triangle with two sides known. Find the third side, using Pythagoras' Theorem.

Calculations in Context

Here is a diagram showing the front of a tent supported by two guyropes, one on each side. The tent is 3 m high and 4 m across. Each guyrope is 7.8 m.

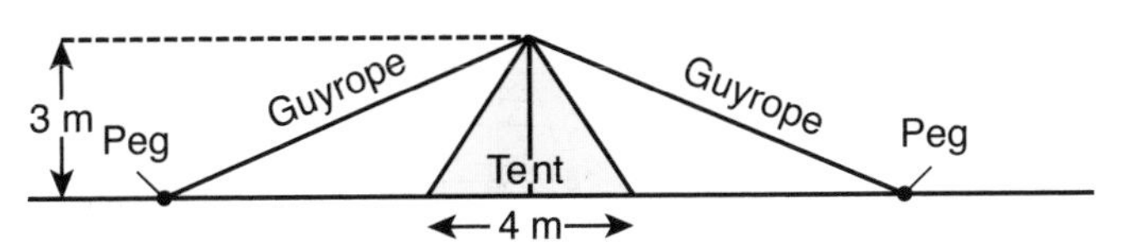

How far out from the edge of the tent is the guyrope peg?

Solution:

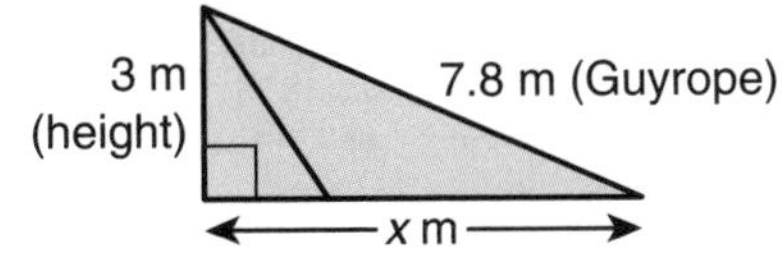

This diagram shows a right-angled triangle with two of the lengths known. Pythagoras' Theorem gives: $x^2 = 7.8^2 - 3^2 = 51.84$ (small square)

so $x = \sqrt{51.84} = 7.2$ m

Here is a diagram with lengths along the ground:

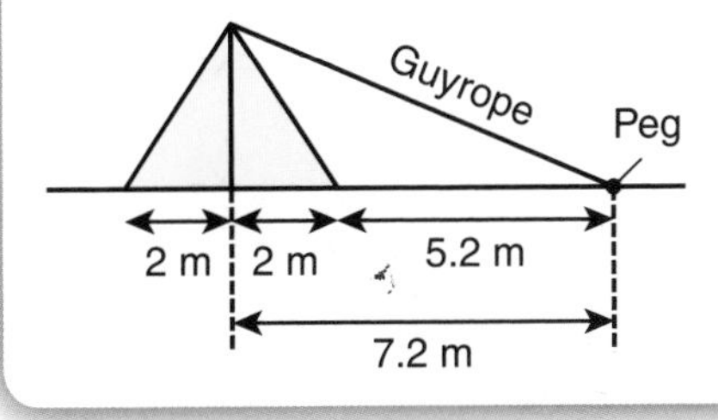

The peg is $7.2 - 2 = 5.2$ m out from the edge of the tent

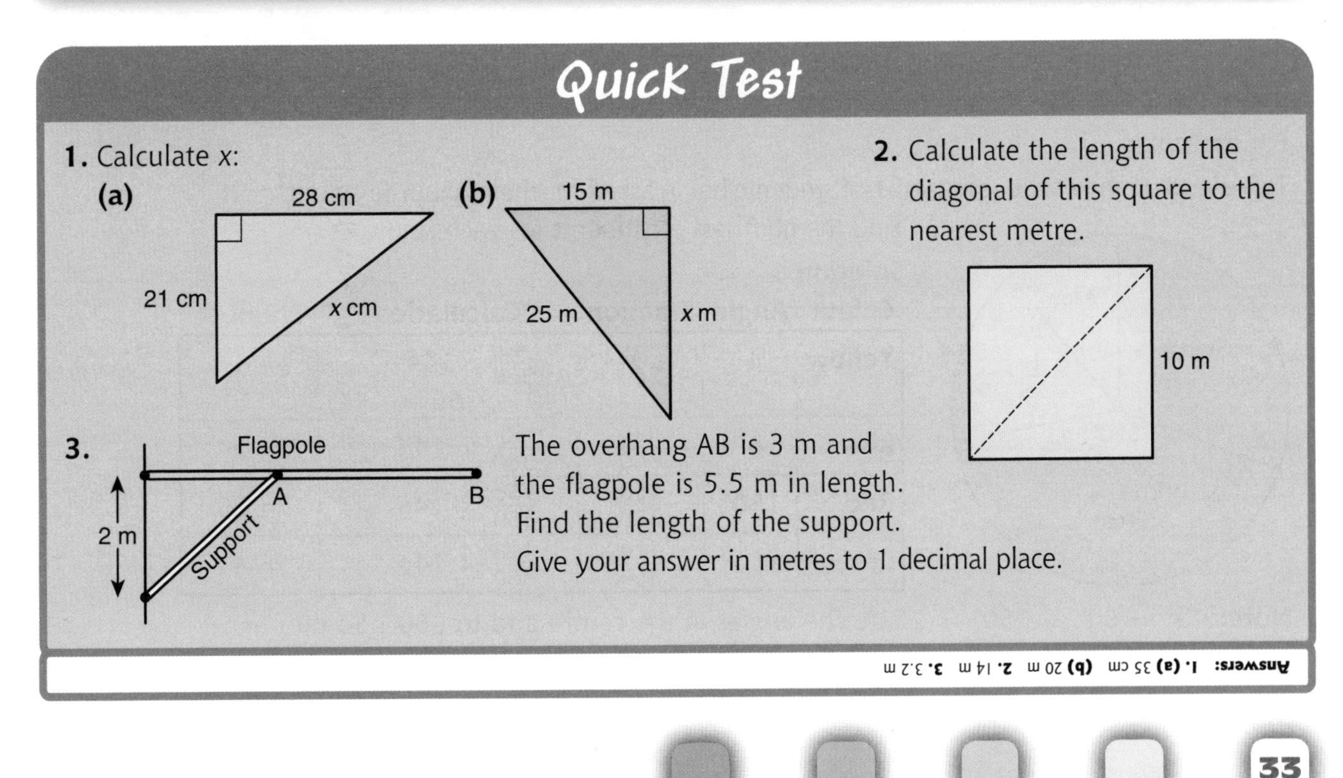

Quick Test

1. Calculate x:

(a)

(b)

2. Calculate the length of the diagonal of this square to the nearest metre.

3.

The overhang AB is 3 m and the flagpole is 5.5 m in length. Find the length of the support. Give your answer in metres to 1 decimal place.

Answers: 1. (a) 35 cm **(b)** 20 m **2.** 14 m **3.** 3.2 m

Data Displays

Bar and Line Graphs

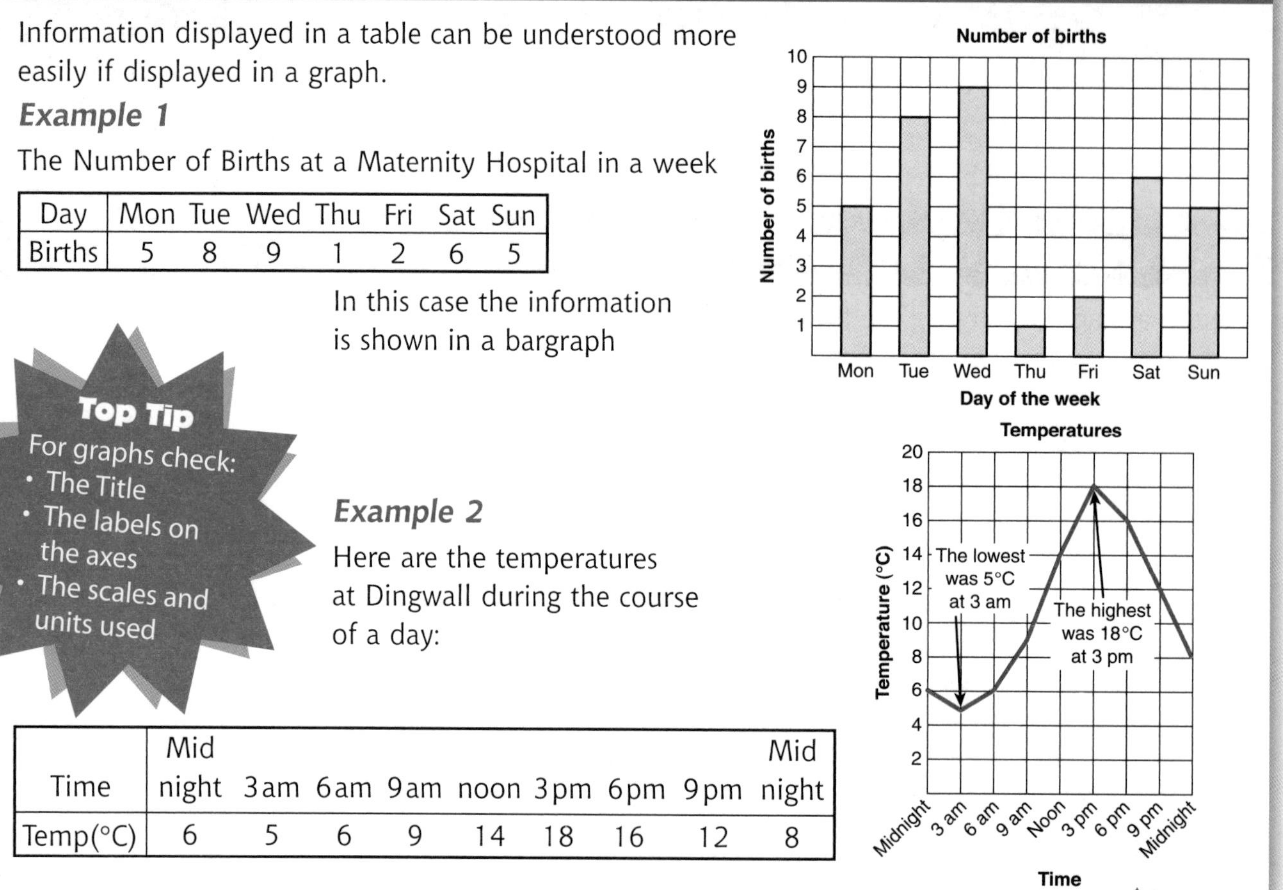

Information displayed in a table can be understood more easily if displayed in a graph.

Example 1

The Number of Births at a Maternity Hospital in a week

Day	Mon	Tue	Wed	Thu	Fri	Sat	Sun
Births	5	8	9	1	2	6	5

In this case the information is shown in a bargraph

Top Tip
For graphs check:
- The Title
- The labels on the axes
- The scales and units used

Example 2

Here are the temperatures at Dingwall during the course of a day:

Time	Mid night	3am	6am	9am	noon	3pm	6pm	9pm	Mid night
Temp(°C)	6	5	6	9	14	18	16	12	8

Pie Charts

Top Tip
For Pie Chart calculations find the fraction of 360° for the slice you are interested in.

A Pie Chart displays information as 'slices' of a pie.
The angles at the centre of the pie are very important when doing calculations using a pie chart

For example:

This pie chart shows the results of 144 students being asked for their favourite colour.

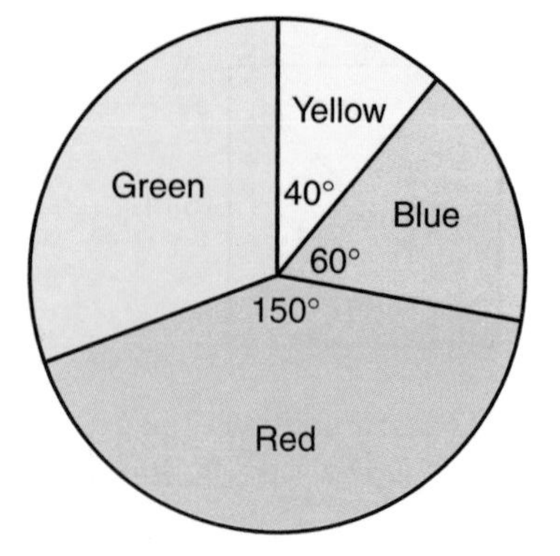

Find the numbers of students for each colour.

Solution:

Colour	Angle	Fraction	Calculation	Number
Yellow	40°	$\frac{40}{360}$	$\frac{40}{360} \times 144$ $(40 \div 360 \times 144)$	= 16
Blue	60°	$\frac{60}{360}$	$\frac{60}{360} \times 144$	= 24
Red	150°	$\frac{150}{360}$	$\frac{150}{360} \times 144$	= 60
Green	110°	$\frac{110}{360}$	$\frac{110}{360} \times 144$	= 44

Note: 40° + 60° + 150° = 250°. The angles at the centre add to 360°. So the angle for Green is 360° – 250° = 110°

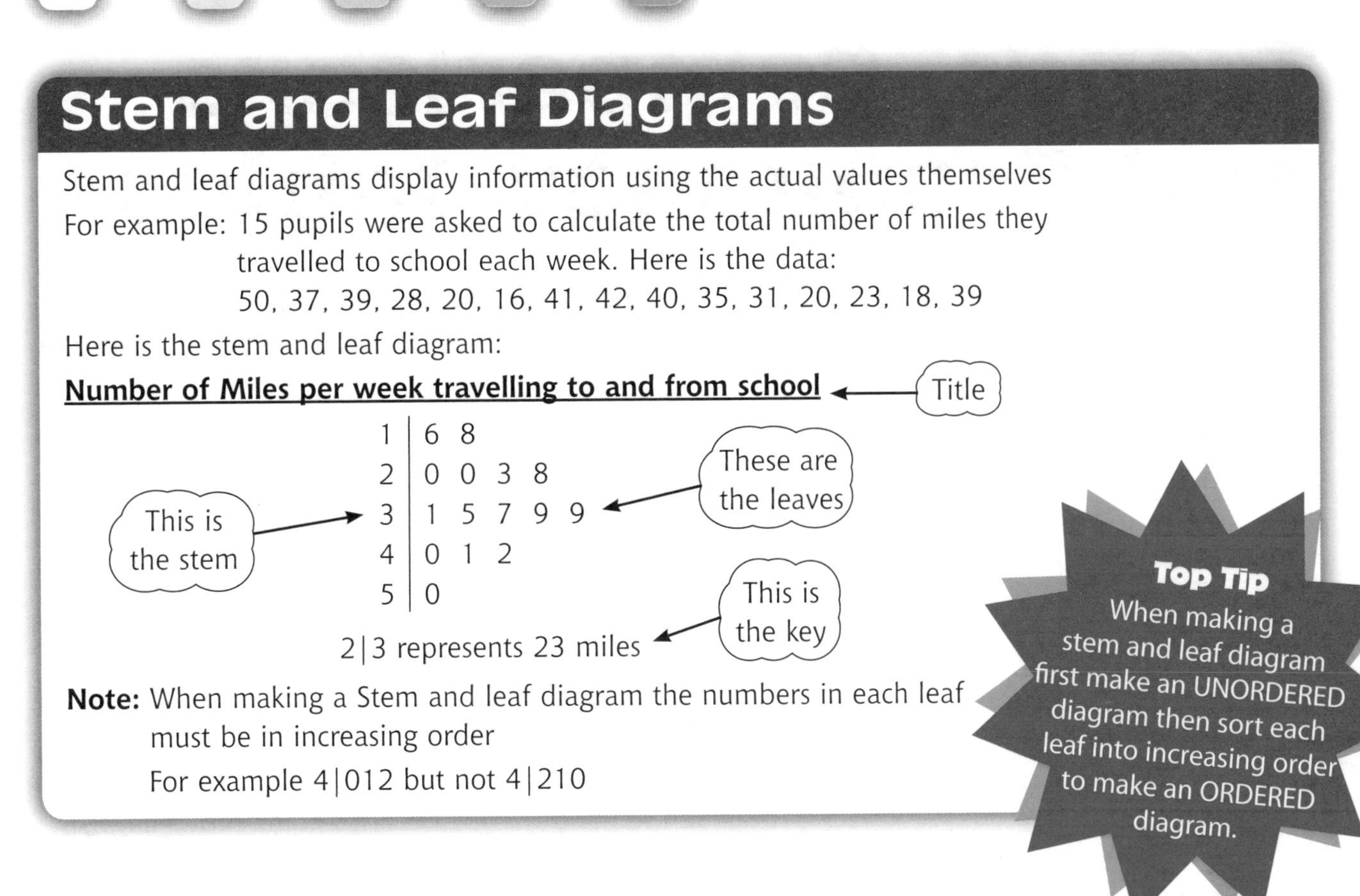

Stem and Leaf Diagrams

Stem and leaf diagrams display information using the actual values themselves

For example: 15 pupils were asked to calculate the total number of miles they travelled to school each week. Here is the data:
50, 37, 39, 28, 20, 16, 41, 42, 40, 35, 31, 20, 23, 18, 39

Here is the stem and leaf diagram:

Number of Miles per week travelling to and from school ← Title

Stem	Leaves
1	6 8
2	0 0 3 8
3	1 5 7 9 9
4	0 1 2
5	0

This is the stem (→ 3)
These are the leaves (→ 1 5 7 9 9)

2|3 represents 23 miles ← This is the key

Note: When making a Stem and leaf diagram the numbers in each leaf must be in increasing order

For example 4|012 but not 4|210

Top Tip
When making a stem and leaf diagram first make an UNORDERED diagram then sort each leaf into increasing order to make an ORDERED diagram.

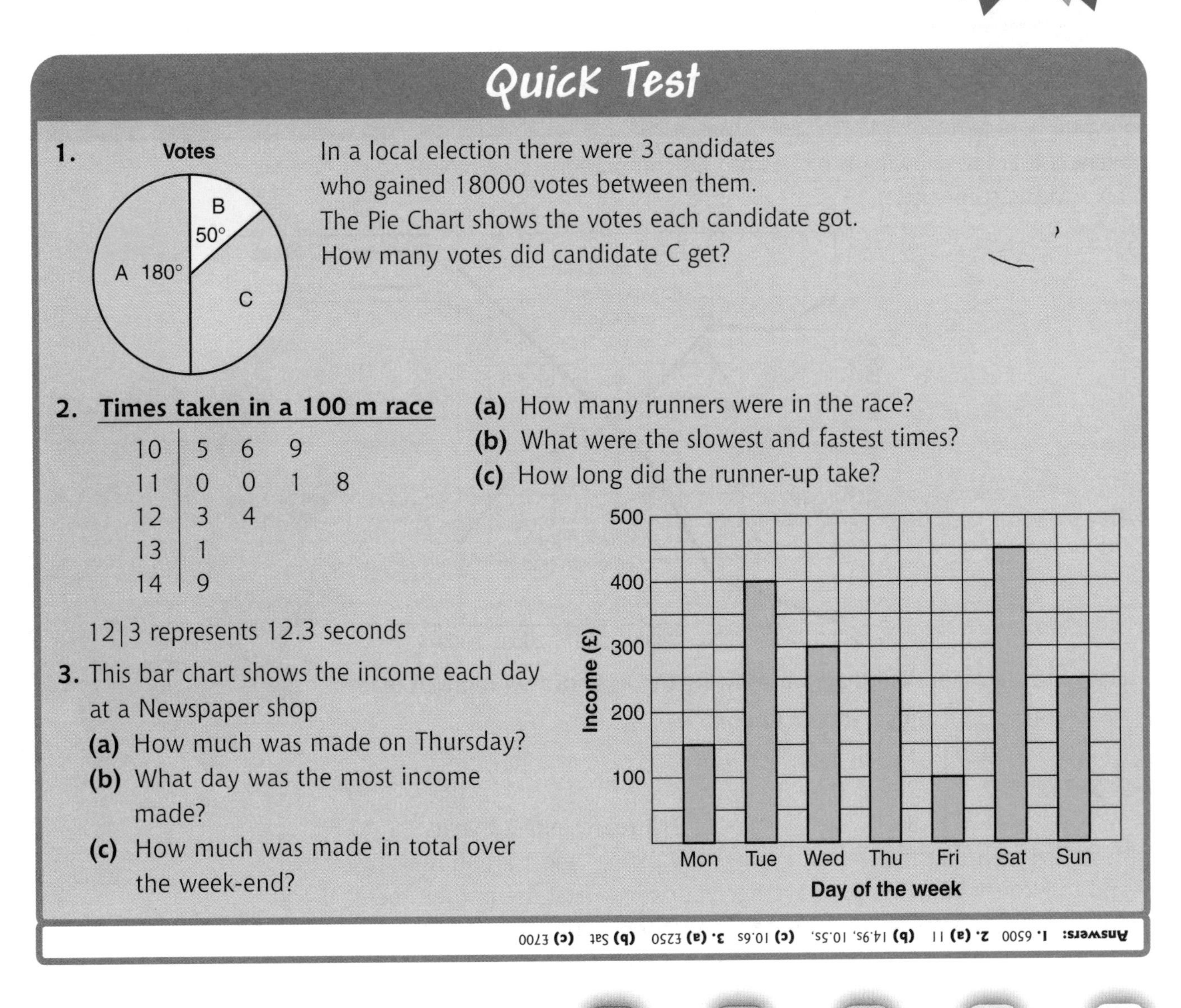

Quick Test

1. In a local election there were 3 candidates who gained 18000 votes between them. The Pie Chart shows the votes each candidate got. How many votes did candidate C get?

2. **Times taken in a 100 m race**

Stem	Leaves
10	5 6 9
11	0 0 1 8
12	3 4
13	1
14	9

12|3 represents 12.3 seconds

(a) How many runners were in the race?
(b) What were the slowest and fastest times?
(c) How long did the runner-up take?

3. This bar chart shows the income each day at a Newspaper shop
(a) How much was made on Thursday?
(b) What day was the most income made?
(c) How much was made in total over the week-end?

Answers: **1.** 6500 **2. (a)** 11 **(b)** 14.9s, 10.5s. **(c)** 10.6s **3. (a)** £250 **(b)** Sat **(c)** £700

Trends, Comparisons and Frequency Tables

Trends

Here are some examples of graphs that show trends:

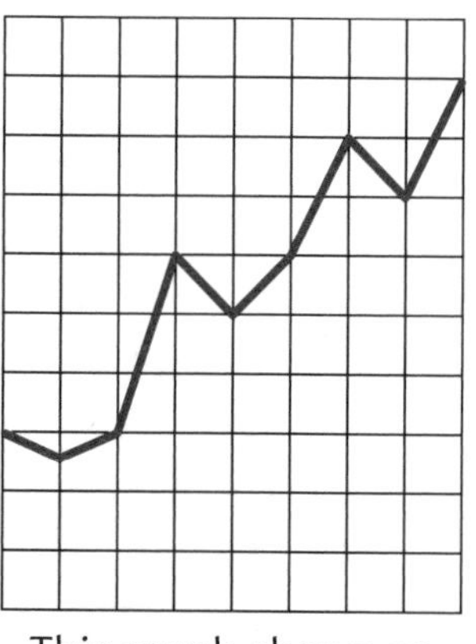

This graph shows an upward trend. In the long run values are increasing.

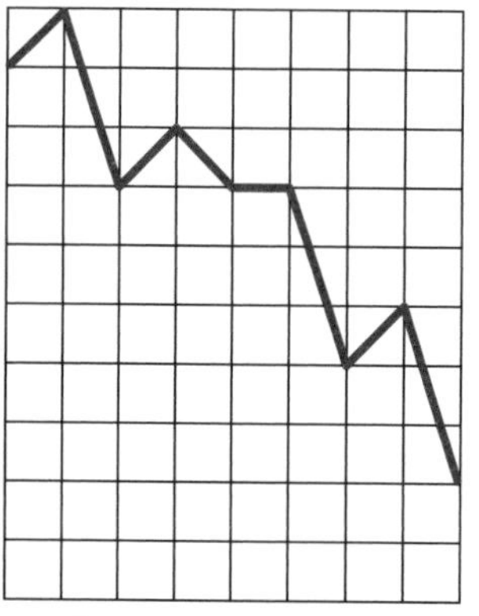

This graph shows a downward trend. In the long run values are decreasing.

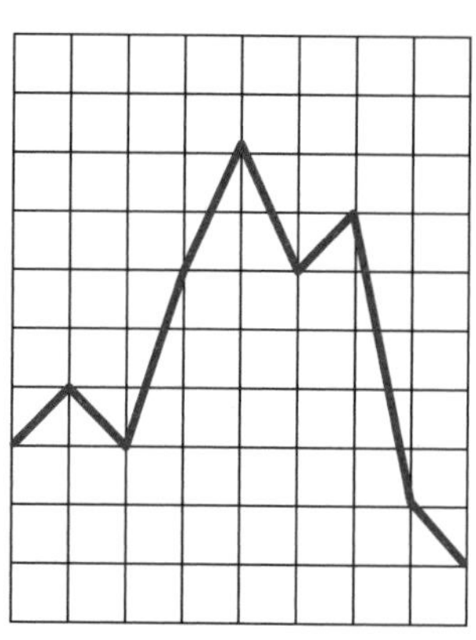

This graph shows an upward trend followed by a downward trend.

Comparisons

Here is a graph showing the sales of CDs compared to downloads sold by a Music company:

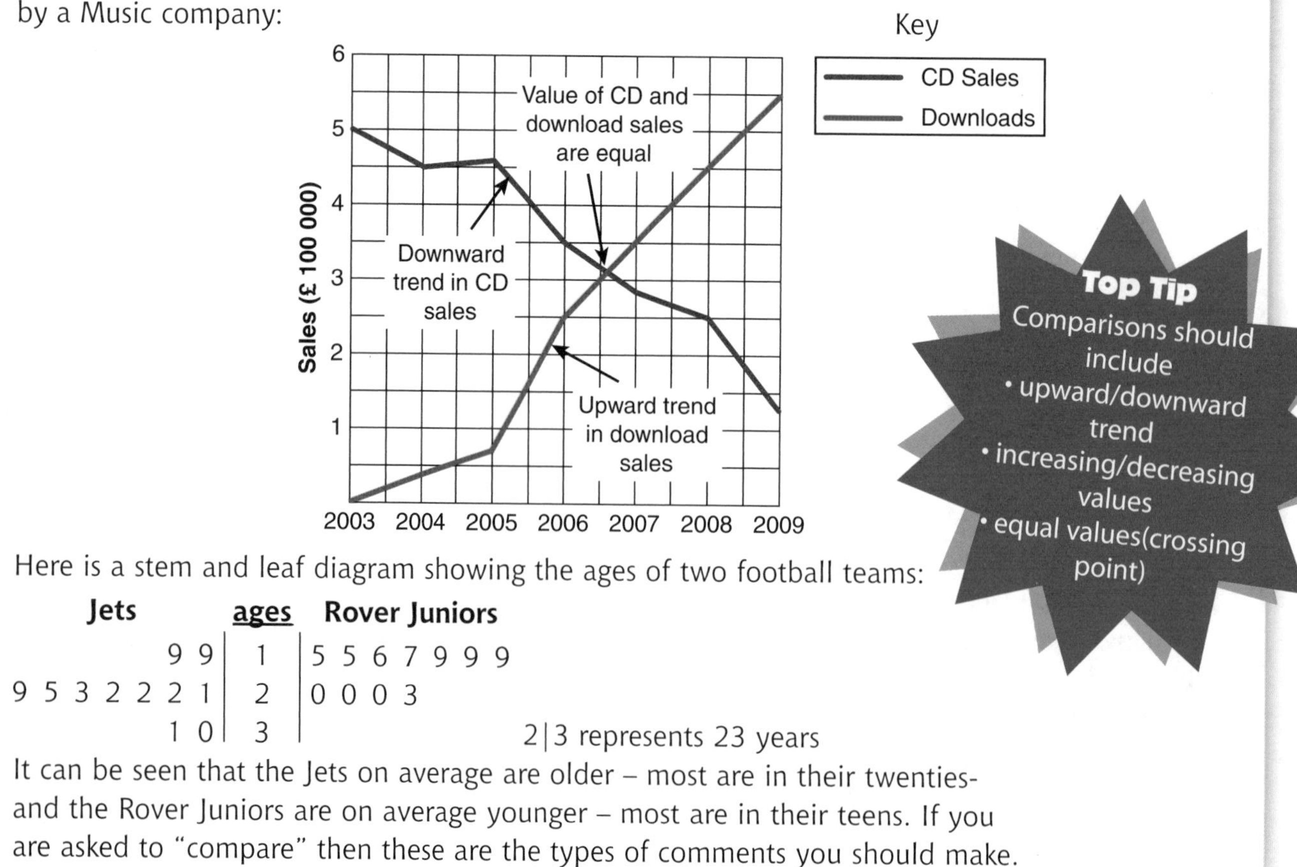

Here is a stem and leaf diagram showing the ages of two football teams:

Jets	ages	Rover Juniors
9 9	1	5 5 6 7 9 9 9
9 5 3 2 2 2 1	2	0 0 0 3
1 0	3	

2|3 represents 23 years

It can be seen that the Jets on average are older – most are in their twenties- and the Rover Juniors are on average younger – most are in their teens. If you are asked to "compare" then these are the types of comments you should make.

Frequency Tables

Information collected, for instance in a survey, can be organised neatly into a frequency table.

Here is an example: A group of teenagers were asked how many times they had been to the cinema in the last month. The results were:

Data (information)

4 0 1 4 0 3 4
0 5 4 0 0 4 1
4 4 3 4 0 0 4
5 3 4 4

When organising this data you should work through the numbers row by row making tally marks in the middle column of the table →

Frequency Table

No of visits	Tally	Frequency
0	𝍸 ll	7
1	ll	2
2		0
3	lll	3
4	𝍸 𝍸 l	11
5	ll	2

Total = 25

A frequency of 11 next to 4 visits means that 11 of the teenagers visited the cinema 4 times in the last month.

Top Tip
Always check the total frequency matches the number of values given.

Quick Test

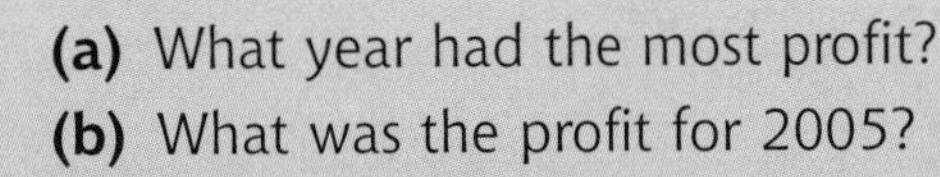

1. Profit made by a Garden Furniture Company

Profit (£ 1000) against Year (2005, 2006, 2007, 2008, 2009)

(a) What year had the most profit?
(b) What was the profit for 2005?
(c) Describe the overall trend in profits for the years shown

2. 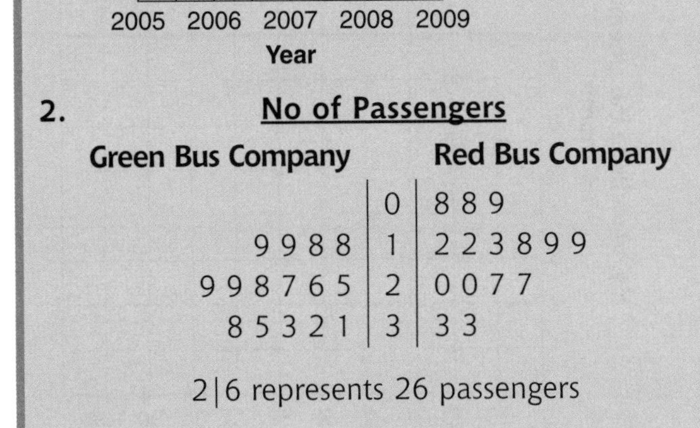

No of Passengers

Green Bus Company		Red Bus Company
	0	8 8 9
9 9 8 8	1	2 2 3 8 9 9
9 9 8 7 6 5	2	0 0 7 7
8 5 3 2 1	3	3 3

2|6 represents 26 passengers

15 buses were surveyed from two bus companies on the same route during the course of a week and the number of passengers recorded. The results are shown in the stem-and-leaf diagram.

(a) What was the fewest number of passengers recorded?
(b) How many buses had more than 30 passengers?
(c) Compare the number of passengers for the two companies. Comment on any differences.

Answers: **1. (a)** 2009 **(b)** £3000 **(c)** upward **2. (a)** 8 **(b)** 7 **(c)** on average, Green has more passengers

Scattergraphs

Making a scattergraph

"Students who do well at Maths generally do well at Physics". To check this statement the test marks of 15 students were recorded (to nearest 10%):

Student:	A	B	C	D	E	F	G	H	I	J	K	L	M	N	O
Maths %:	50	50	30	80	80	20	60	70	90	40	90	80	40	70	60
Physics %:	20	40	20	70	80	30	90	90	90	80	80	100	50	50	70

The best way of checking the truth of the statement is to display these marks in a scatter graph. Each pair of marks is plotted as one point.

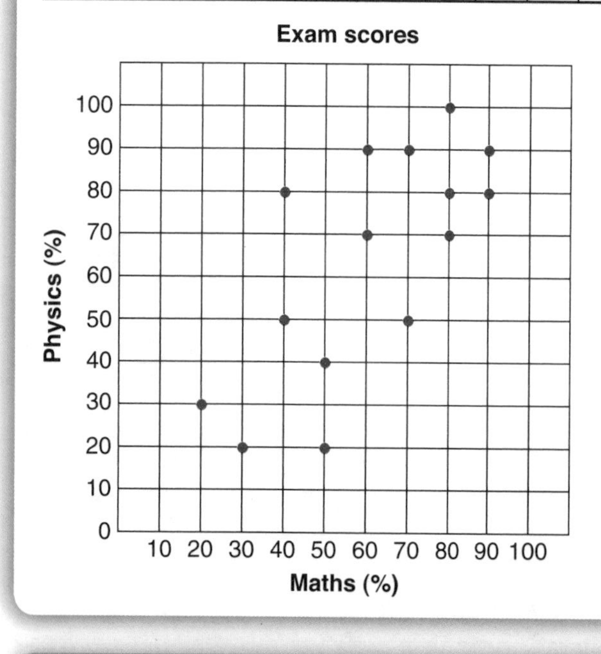

For example Student C has scores 30% and 20% in Maths and Physics and is shown on the scattergraph as point (30, 20).

The scattergraph shows that students with higher Maths scores in general also have higher Physics scores.

This is called a POSITIVE CORRELATION between the Maths and Physics scores.

Top Tip
Two values give one point on a scattergraph.

Correlation

You should know these two types of correlation

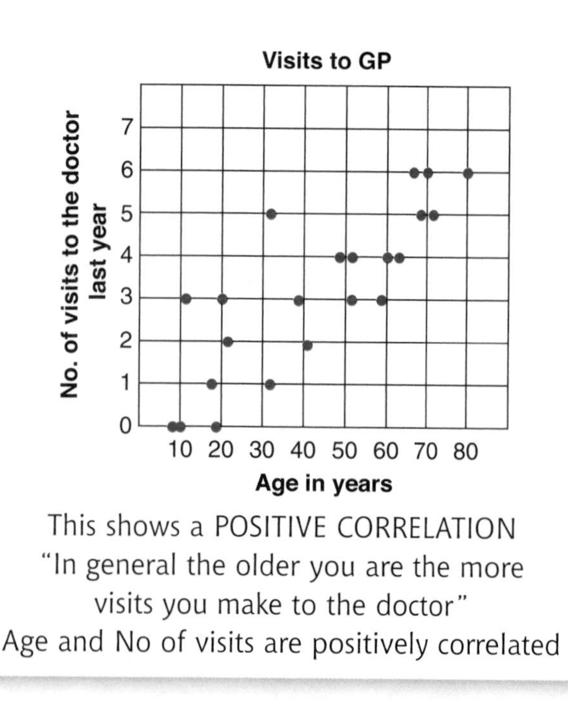

This shows a POSITIVE CORRELATION
"In general the older you are the more visits you make to the doctor"
Age and No of visits are positively correlated

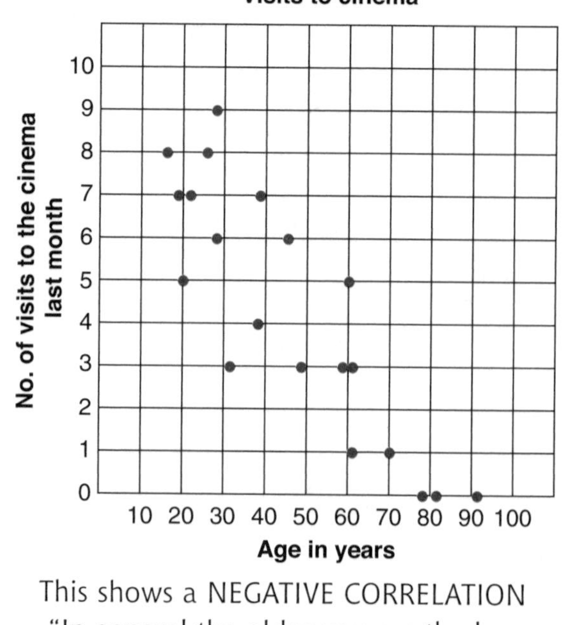

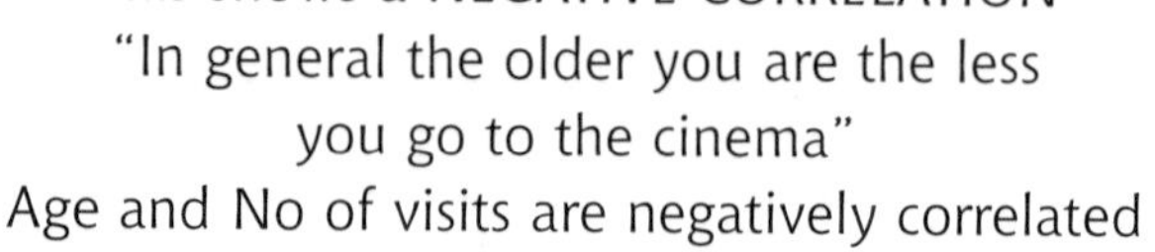

This shows a NEGATIVE CORRELATION
"In general the older you are the less you go to the cinema"
Age and No of visits are negatively correlated

The Line of Best Fit

The points on a scattergraph that shows positive or negative correlation tend to lie along a line. This line is called the LINE OF BEST FIT.

When drawing a line of best fit try to get roughly the same number of points above the line as below.

You are trying to draw the line through the 'middle' of the points while following the trend (up or down).

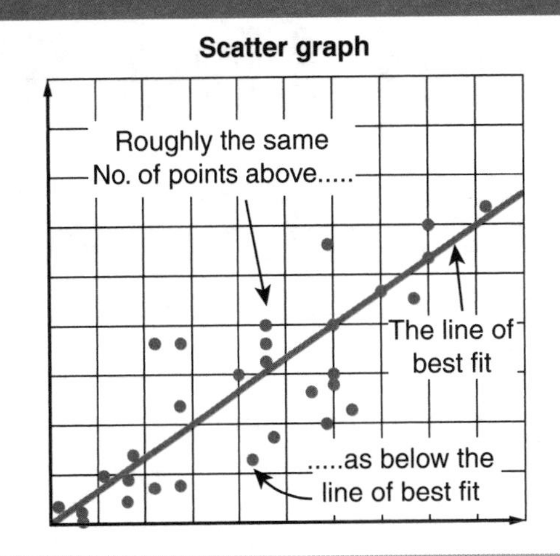

Using the Line of Best Fit

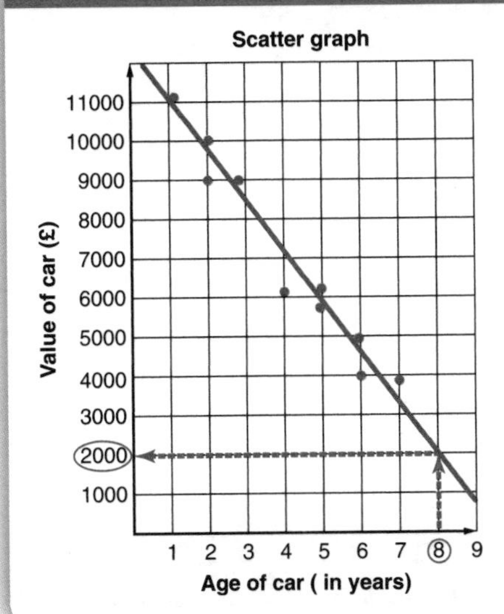

Here is an example:

A line of best fit has been drawn on this scattergraph. It can be used to estimate values.

As the age of a car increases its value gets less. This is negative correlation and the line of best fit slopes downwards.

The line passes through the point (8, 2000)

This means an 8-year-old car is likely to be worth £2000.

Top Tip

When drawing a line of best fit – ignore the few points far away from the rest.

Quick Test

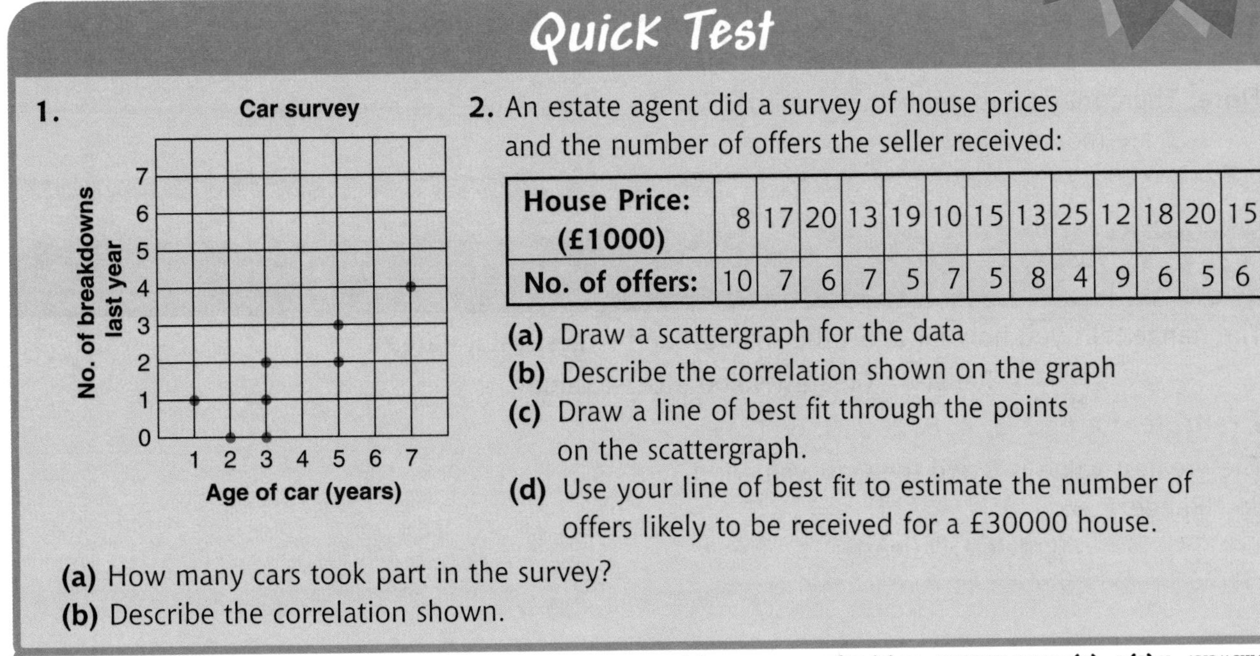

1.

(a) How many cars took part in the survey?

(b) Describe the correlation shown.

2. An estate agent did a survey of house prices and the number of offers the seller received:

House Price: (£1000)	8	17	20	13	19	10	15	13	25	12	18	20	15
No. of offers:	10	7	6	7	5	7	5	8	4	9	6	5	6

(a) Draw a scattergraph for the data

(b) Describe the correlation shown on the graph

(c) Draw a line of best fit through the points on the scattergraph.

(d) Use your line of best fit to estimate the number of offers likely to be received for a £30000 house.

Answers: 1. (a) 8 **(b)** Positive correlation **2. (b)** Negative correlation **(d)** 2 or 3

Types of Average and the Range

Mean

This average is calculated using the formula

$$\text{Mean} = \frac{\text{Total of all values}}{\text{Number of values}}$$ (This is an addition followed by a division)

Example: The mean of 6, 2, 3, 4, 8, 2, 2, 5, 6, 1 is given by

$$\text{Mean} = \frac{6+2+3+4+8+2+2+5+6+1}{10}$$ ← total of values

$$= \frac{39}{10} = 3.9$$ ← there are 10 values

Top Tip
MEAN: Add and Divide
MODE: Most frequent
MEDIAN: Middle in order

Median

To find this average:

Step 1: Sort the values into order – smallest to largest.

Step 2: Find the middle value or calculate the number half-way between the middle two values.

Example: For 6, 2, 3, 4, 8, 2, 2, 5, 6, 1

Step 1: In increasing order 1, 2, 2, 2, 3, 4, 5, 6, 6, 8

Step 2: There are two middle values, 3 and 4 so 3.5 is the Median.

Mode

For this average you pick the most frequent value.

Example: for 6, 2, 3, 4, 8, 2, 2, 5, 6, 1 The mode is 2 since there are more 2's than any other value.

Note: There may be more than one mode. For the values 2, 2, 3, 3 both 2 and 3 are modal values.

Range

The Range tells you how far apart the smallest and largest values are:

Range = Greatest value – Least value

Example: for 6, 2, 3, 4, 8, 2, 2, 5, 6, 1

The greatest value is 8 and the least value is 1

So Range = 8 (greatest) – 1 (least) = 7

Averages and Range from a Frequency Table

Example: During the course of one week Bryn recorded the lengths of all his mobile phone calls (to the nearest minute).

Length of call (minutes)	Frequency	Length x Frequency
1	14	$1 \times 14 = 14$
2	8	$2 \times 8 = 16$
3	5	$3 \times 5 = 15$
4	3	$4 \times 3 = 12$
	Total = 30	Total = 57

He made 30 calls in total

This is the total length of all his calls

Remember:

$$\text{Mean} = \frac{\text{Total of all values}}{\text{Number of values}}$$

In this case

$$\text{Mean} = \frac{\text{Total length of all calls}}{\text{Number of calls}}$$

$$= \frac{57}{30} = 1.9 \text{ minutes}$$

The Mode is the most frequent – in this case 14 of his calls were 1 minute calls – the most frequent – so 1 minute is the Mode.

Also: Range = 4 min (greatest) – 1 min (least) = 3 minutes.

Top Tip

Values	Frequency	Value x Frequency
⋮	⋮	⋮
	Total A	Total B

$\text{Mean} = \frac{B}{A}$

Finding the Mode from a Stem-and-Leaf Diagram

Example:

```
1 | 2
2 | 0 4 6 8 9
3 | (1) 2 3 6
4 | 2 5 5
```

31 is the Median

2|6 represents 26

In this Stem-and-Leaf diagram there are 13 values: 12, 20, 24, ... 42, 45, 45

The middle value is 31. There are 6 values on the leaves below 31 and 6 values on the leaves above 31.

Quick Test

1. For the values: 2, 8, 3, 9, 4, 12, 3, 8, 9, 8, 3, 4, 3, 5, 3
Calculate: **(a)** The Mean **(b)** The Mode **(c)** The Median **(d)** The Range

2.

No of visits	Frequency	No of visits × Frequency
1	7	
2	7	
3	5	
4	1	
	Total =	Total =

Complete this frequency table and calculate the mean number of visits.

3. Find the median value for this Stem-and-Leaf diagram:

```
2 | 0 1 2 3 7 7 8 9
3 | 0 1 2 4 6
4 | 8 9 9
5 | 0 1
```

3|4 represents 34

Answers: 1. (a) 5.6 **(b)** 3 **(c)** 4 **(d)** 10 **2.** 20, 40, mean = 2 **3.** 30.5

Probability and Interpreting Statistics

Comparing two sets of data

10 girls and 10 boys at a school were asked how many minutes they spend on average each evening doing homework.

The results were:

Boys	15	45	10	25	0	75	15	0	40	0
Girls	40	60	65	30	65	35	70	40	60	30

The Median for the boys is 15 minutes. The Median for the girls is 50 minutes.

Comparing the medians you can say: "On average the girls spent more time (35 minutes) on homework than the boys."

The Range for the boys is 75 minutes. The Range for the girls is 40 minutes.

Comparing the ranges you can say:
"There is more variation in the times spent by the boys on homework compared to the times spent by the girls."

Note: A greater range means the values are more spread out.

Which Average?

Here are the attendance figures for 8 football matches:

6000, 9500, 4500, 4000, 6500, 80000, 5500, 4000

The Mean is $\frac{6000 + 9500 + 4500 + 4000 + 6500 + 80000 + 5500 + 4000}{8}$

$= 15000$

4000, 4000, 4500, 5500, 6000, 6500, 9500, 80000

The Median is $\frac{5500+6000}{2} = 5750$

Twice the attendance was 4000. The Mode is 4000.

In this case the average that gives the true picture is the median. The 80000 attendance was unusual and has a big effect on the mean. The median is unaffected by this large attendance.

The mode is not realistic as most of the attendance figures are all different and it was just chance that two were the same.

Top Tip
Use the Range for how 'spread out' the values are.

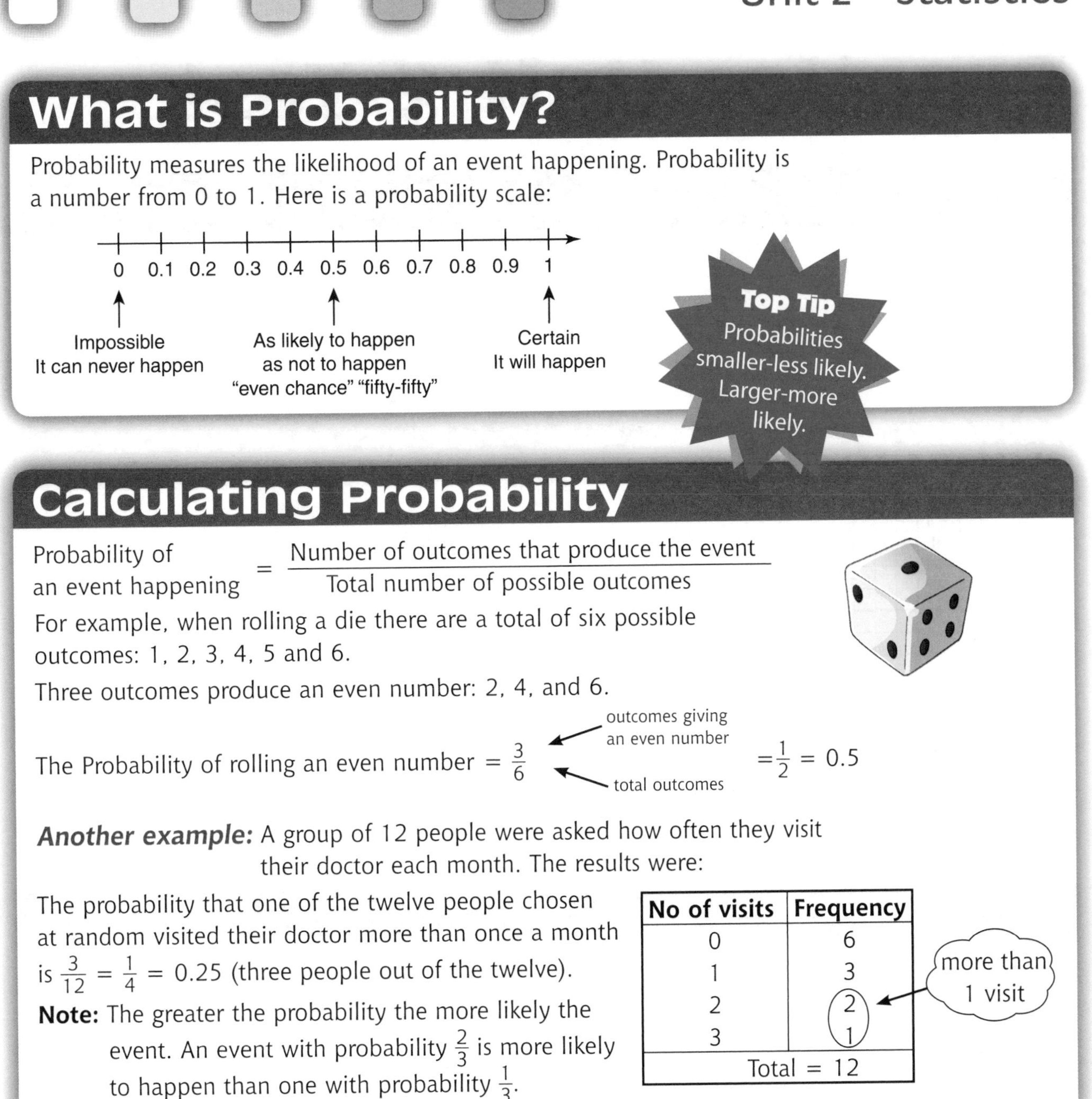

What is Probability?

Probability measures the likelihood of an event happening. Probability is a number from 0 to 1. Here is a probability scale:

Top Tip
Probabilities smaller-less likely. Larger-more likely.

Calculating Probability

$$\text{Probability of an event happening} = \frac{\text{Number of outcomes that produce the event}}{\text{Total number of possible outcomes}}$$

For example, when rolling a die there are a total of six possible outcomes: 1, 2, 3, 4, 5 and 6.

Three outcomes produce an even number: 2, 4, and 6.

The Probability of rolling an even number $= \frac{3}{6}$ (3 = outcomes giving an even number; 6 = total outcomes) $= \frac{1}{2} = 0.5$

Another example: A group of 12 people were asked how often they visit their doctor each month. The results were:

No of visits	Frequency
0	6
1	3
2	2
3	1
Total = 12	

(2 and 1: more than 1 visit)

The probability that one of the twelve people chosen at random visited their doctor more than once a month is $\frac{3}{12} = \frac{1}{4} = 0.25$ (three people out of the twelve).

Note: The greater the probability the more likely the event. An event with probability $\frac{2}{3}$ is more likely to happen than one with probability $\frac{1}{3}$.

Quick Test

1. Aberdeen temperatures during the week had a mean of 7°C with a range of 15° C. London during the same week had a mean of 12° C and a range of 6° C. Make two comments comparing the temperatures in Aberdeen and London.

2. 

This 'Scrabble' word scores 12.

If these 6 letters are put in a bag and one chosen at random, what is the probability it is:

(a) a vowel **(b)** has a score of 1 **(c)** has a score of 4?

Answers: 1. On average, London had higher temperatures; Aberdeen temperatures had a greater variation **2. (a)** 1/3 **(b)** 2/3 **(c)** 0

Practice Unit 2 Test

Formulae

You will be given this formula (among others) for your Unit 2 assessment:

Theorem of Pythagoras:

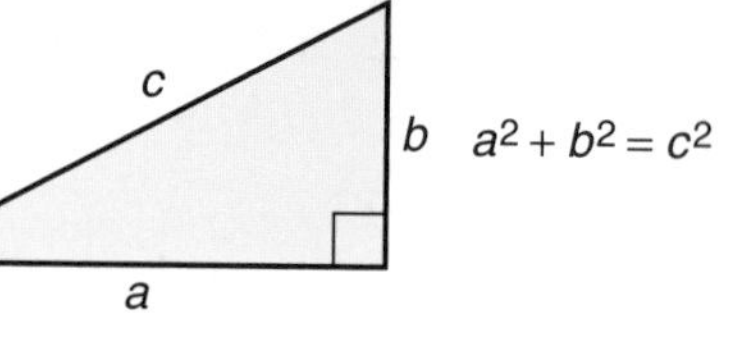

Outcome 1

1. **(a)** State the coordinates of the points A and B.

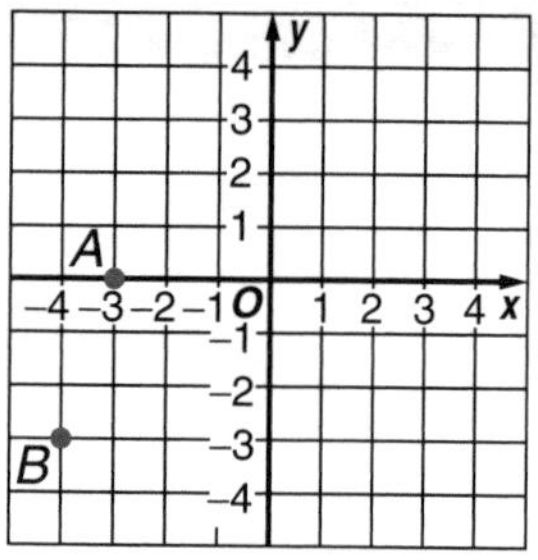

(b) Plot the points C(0, −2) and D(3, −2) on the diagram

2. (a) Calculate the total for row 2

(b) Calculate the total for row 3

(c) By how much is the total for row 4 greater than the total for row 1?

				row total
row 1:	3	−2	−8	−7
row 2:	5	6	−4	
row 3:	−8	2	0	
row 4:	4	0	−1	3

Outcome 2

3. The graph shows Peter's journey to Edinburgh. He stopped for a break on the way.

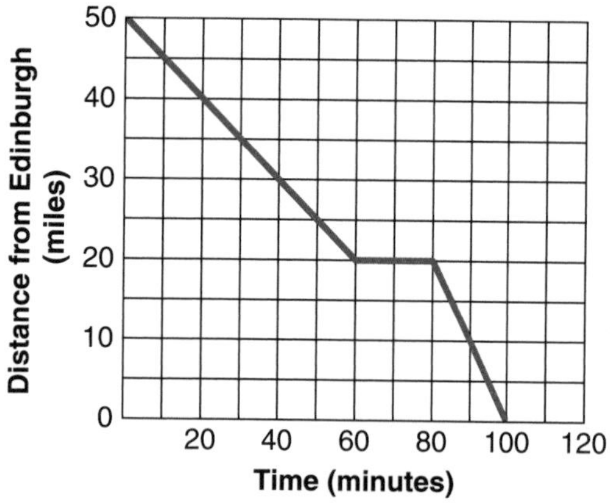

(a) How far did he travel before his break?

(b) How long was his break?

(c) How can you tell from the graph that his average speed was faster after his break than before?

4. If I travel at 45 km per hour for 3 hours how far have I travelled?

5. If I travel 3640 miles at an average speed of 80 miles per hour, how long do I take?

Outcome 3

6. The positions of three pupils in a classroom form a right-angled triangle.

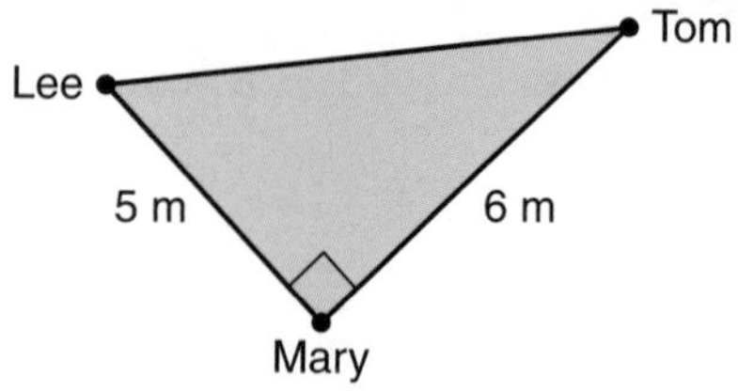

Calculate the distance from Lee to Tom.

Outcome 4

7. 12 Pupils sat a Maths test. Here are their percentage scores:
70, 63, 59, 41, 68, 64, 58, 60, 82, 71, 62, 65

(a) Write down the maximum and minimum percentages scored.

(b) Complete this stem-and-leaf diagram for the data. Remember to include a key

4	
5	
6	
7	
8	

(c) Comment on what the stem-and-leaf diagram shows.

8.

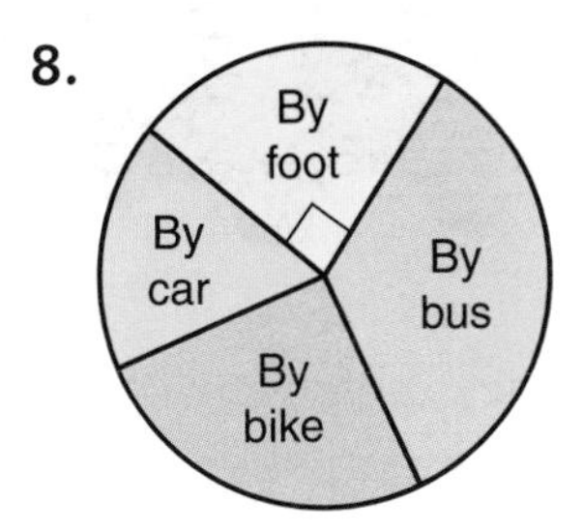

80 pupils were asked how they travelled to school.
The results are shown in t.his pie-chart.
How many pupils walked to school?

9. Twenty pupils were asked "how many brothers and sisters do you have?"
Here are their replies: 0, 1, 2, 1, 0, 3, 1, 1, 0, 1, 2, 1, 4, 0, 1, 4, 0, 1, 1, 0

(a) Complete this frequency table:

Score	Tally	Frequency
⋮	⋮	⋮

(b) A teacher said "Most of these pupils have two or more brothers and sisters". Comment on this statement using your completed frequency table.

10. Nine students were asked to memorise a string of digits and then a string of letters. The table shows how many of each they memorised:

Student:	A	B	C	D	E	F	G	H	I
No of digits:	8	10	6	8	10	6	4	9	5
No of letters:	7	10	4	6	8	3	2	8	4

(a) Draw a Scattergraph

(b) Draw a best-fitting line

(c) Use this line to estimate the number of letters memorised by someone who memorised 12 digits.

Outcome 5

11. 12 kg, 8 kg, 7 kg, 15 kg, 6 kg, 9 kg, 7 kg, 6 kg, 7 kg, 11 kg. For these weights calculate the: **(a)** Mean **(b)** Mode **(c)** Median **(d)** Range

12. Yvonne chose a letter at random from the word "LECKIE". What is the probability that it was the letter E?

Preparation for Assessment

Practice Unit 2 Exam Questions

1. Tim leaves Edinburgh at 21:55 on the night bus arriving in London at 07:45 the next morning.

How long was his journey?

2.

Number of Journeys per week	Frequency	Number of Journeys × Frequency
0	8	0
1	10	10
2	15	30
3	12	
4	4	
5	1	
	Total = 50	

The Frequency table shows the number of journeys over 50 miles taken each week by 50 employees in a company.

(a) Complete the table and calculate the mean number of journeys per week.

(b) If an employee was chosen at random what is the probability that they made 1 or 2 journeys each week?

3. (a) State the coordinates of A and B.

(b) On the grid plot the point C(0, –2)

(c) Plot the point D so that ABCD is a square and write down the coordinates of the point D.

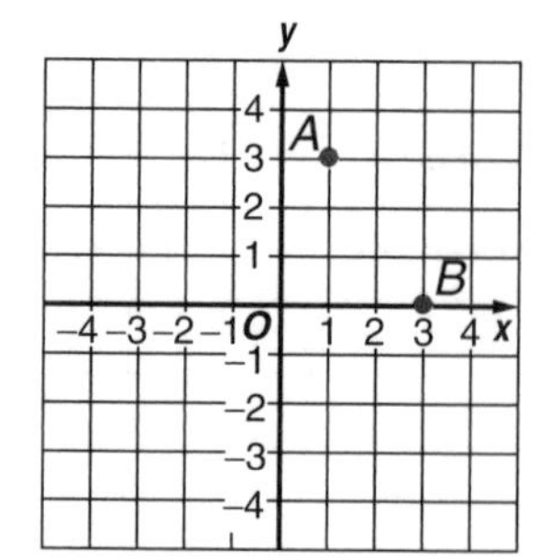

4. 

10 boys were asked how much pocket money they got each week. The results were as follows:

£2, £5, £7, £1, £1, £6, £3, £2, £5, £3

(a) Find the median

(b) Find the Range

(c) 10 girls were asked the same question. For the girls the median was £5 and the range was £3. Make two comments comparing the pocket money received each week by the boys and the girls.

5. Leanne drove 85 miles in $2\frac{1}{2}$ hours.

(a) What was her average speed?

(b) She then drove a further 77 miles in $4\frac{1}{4}$ hours. What was her average speed for the whole journey?

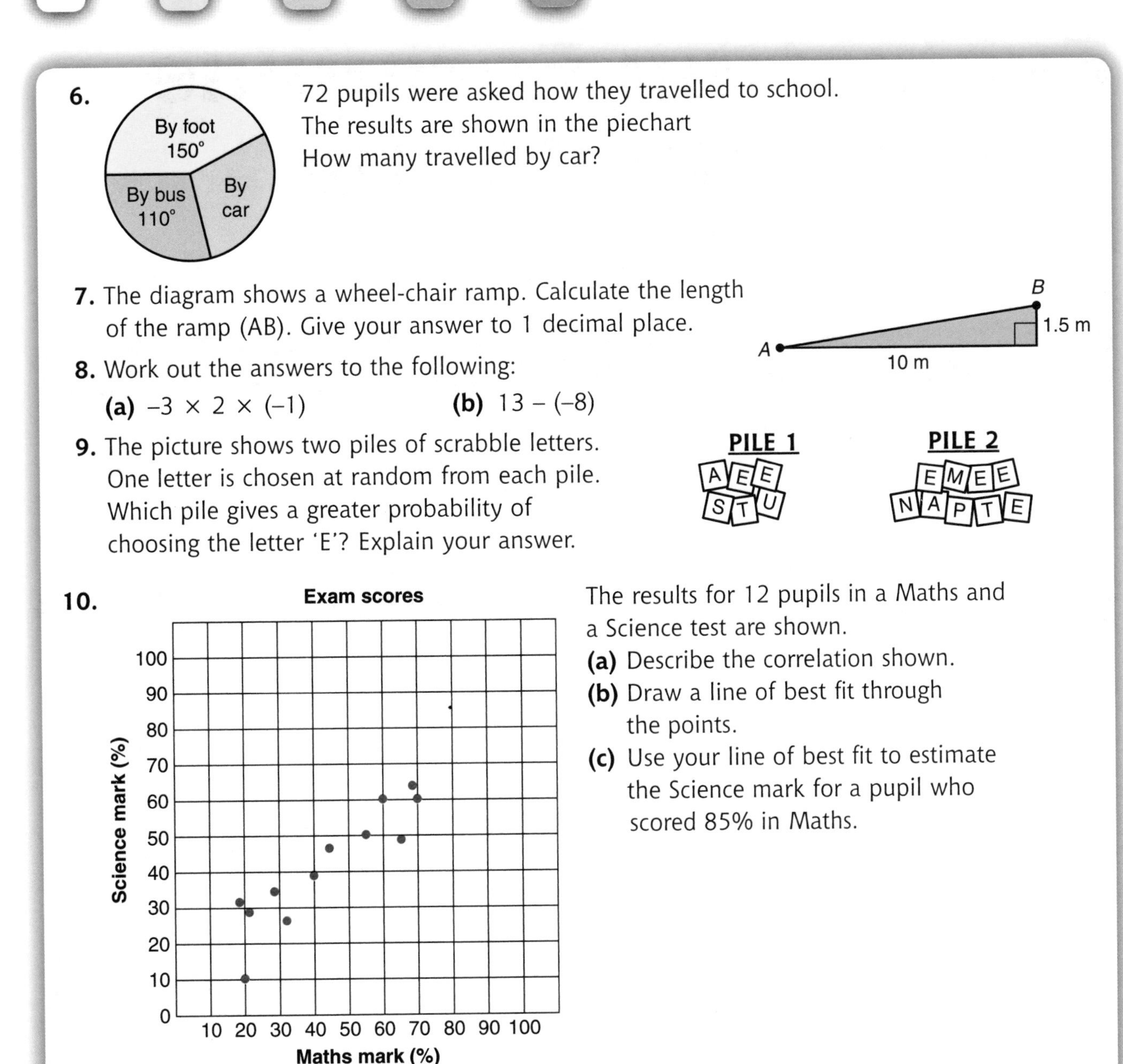

6. 72 pupils were asked how they travelled to school. The results are shown in the piechart
How many travelled by car?

7. The diagram shows a wheel-chair ramp. Calculate the length of the ramp (AB). Give your answer to 1 decimal place.

8. Work out the answers to the following:

(a) $-3 \times 2 \times (-1)$ **(b)** $13 - (-8)$

9. The picture shows two piles of scrabble letters. One letter is chosen at random from each pile. Which pile gives a greater probability of choosing the letter 'E'? Explain your answer.

10. The results for 12 pupils in a Maths and a Science test are shown.

(a) Describe the correlation shown.

(b) Draw a line of best fit through the points.

(c) Use your line of best fit to estimate the Science mark for a pupil who scored 85% in Maths.

11. The picture shows a wooden gate with two diagonal slats meeting in the middle. Calculate the length of the diagonals.

12. The value of six houses in a street are:

£120000, £130000, £95000, £140000, £650000, £110000

(a) Find the mean value

(b) Find the median value

(c) Which average gives a truer picture of the value of the houses in the street? Give a reason for your answer.

Evaluating Formulae and Manipulating Expressions

Basic Notation

$2x$ means	$3a$ means	m^2 means	$\sqrt{n}$ means	$\frac{a}{b}$ means $a \div b$
$2 \times x$	$3 \times a$	$m \times m$	"the square root of n"	"a divided by b"
"two lots of x"	"Three lots of a"	"m squared"	examples $\sqrt{16} = 4$ $\sqrt{9} = 3$	

Evaluating Expressions

Follow this order when finding the value of an expression with letters:

First → Work out calculations in brackets –then→ Work out any squares or square roots –then→ Calculate any multiplications or divisions –then→ Calculate any additions or subtractions

Examples of basic calculations:

if $a = 4$ and $b = 2$

$ab = 4 \times 2 = 8$ $\quad a + b = 4 + 2 = 6$ $\quad b^2 = b \times b = 2 \times 2 = 4$

$\frac{a}{b} = \frac{4}{2} = 2$ $\quad a - b = 4 - 2 = 2$ $\quad \sqrt{a} = \sqrt{4} = 2$

...and some involving different orders:

$10 - ab = 10 - 4 \times 2 = 10 - 8 = 2$ (multiplication first)

$ab^2 = 4 \times 2^2 = 4 \times 4 = 16$ (squaring first)

$(2a - b)^2 = (2 \times 4 - 2)^2 = (8 - 2)^2 = 6^2 = 36$ (brackets first)

$3\sqrt{a}b = 3 \times \sqrt{4} \times 2 = 3 \times 2 \times 2 = 6 \times 2 = 12$ (square root first)

...and an example with negative values:

If $m = -2$ and $n = 3$ then $mn - m = (-2) \times 3 - (-2) = -6 - (-2) = -6 + 2 = -4$

Top Tip

Calculation Order is: Brackets–Squares or Square Roots–Multiplication or Divisions–Additions or Subtraction

Formulae Reminders

Formulae are like recipes – they give a set of instructions for a calculation.

Examples:

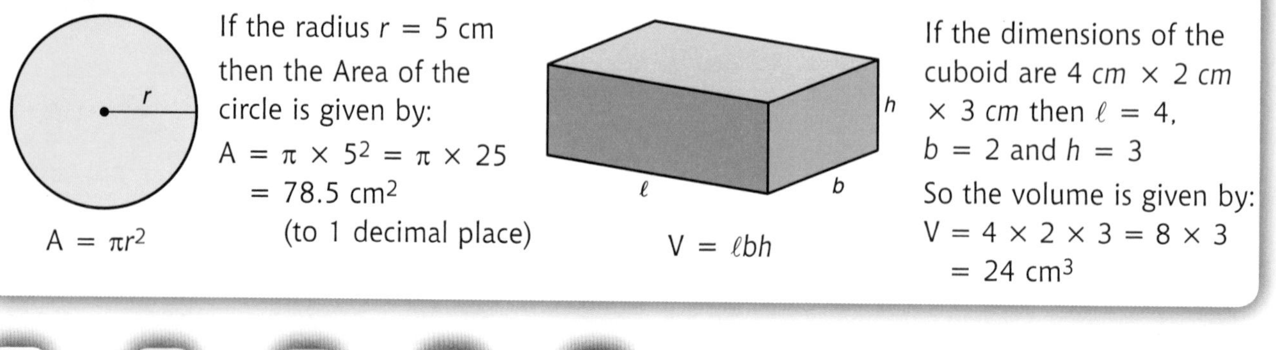

If the radius $r = 5$ cm then the Area of the circle is given by:

$A = \pi \times 5^2 = \pi \times 25$
$= 78.5$ cm^2 (to 1 decimal place)

If the dimensions of the cuboid are $4\ cm \times 2\ cm \times 3\ cm$ then $\ell = 4$, $b = 2$ and $h = 3$

So the volume is given by:

$V = 4 \times 2 \times 3 = 8 \times 3$
$= 24$ cm^3

Evaluating Formulae

Examples:

1. Find Q when $a = 2$, $b = 3$ and $c = 7$ and $Q = c - ab$

Solution: $Q = c - ab = 7 - 2 \times 3 = 7 - 6 = 1$

(multiplication first)

2. Find M when $r = 2.1$, $s = 3.5$ and $n = 1.4$ and $M = \frac{rs}{n^2}$

Solution: $M = \frac{rs}{n^2} = \frac{2.1 \times 3.5}{1.4^2} = \frac{7.35}{1.96} = 3.75$

(calculate top and bottom of fraction then divide)

3. If you are h metres above sea level then the distance, d kilometers, to the horizon is given by the formula: $d = \sqrt{13h}$. Peter is standing at the top of a 30 metre cliff, how far is the horizon?

Solution: $h = 30$ So $d = \sqrt{13h} = \sqrt{13 \times 30} = \sqrt{390} = 19.7$ km (to 1 dec place)

Top Tip
When simplifying add or subtract *like* terms e.g. $3x$ and $2x$

Simplifyling Expressions

Example:

$$2a + 4b + 3a - b = 5a + 3b$$

two lots of 'a' added to three lots of 'a' gives five lots of 'a'

four lots of 'b' take away one lot of 'b' gives three lots of 'b'

This is now a simplified expression

Removing Brackets

With numbers		With letters
$3 \times (4 + 5) = 3 \times 4 + 3 \times 5$	→	$a(b + c) = ab + ac$
($3 \times 9 = 27$) ($12 + 15 = 27$)		
$4 \times (8 - 3) = 4 \times 8 - 4 \times 3$	→	$a(b - c) = ab - ac$
($4 \times 5 = 20$) ($32 - 12 = 20$)		

This process is described as "multiplying out the brackets."

More examples:

1. $5(x - 2)$
$= 5x - 10$

2. $3(4m + 1) + 12$
$= 12m + 3 + 12$
$= 12m + 15$

3. $7(a + 3b) - 2b$
$= 7a + 21b - 2b$
$= 7a + 19b$

Top Tip
$a(b + c) = ab + ac$
both 'b' and 'c' are multiplied by 'a'

Quick Test

1. $A = r - st$ Calculate A when $r = 12$, $s = 2$ and $t = 4$

2. Multiply out and simplify:

(a) $6(x + 2) + 3$ **(b)** $7 + 3(2n - 1)$

3. Find the value of $2ab - c$ when $a = -1$, $b = -3$ and $c = -5$

Answers: 1. 4 **2. (a)** $6x + 15$ **(b)** $4 + 6n$ **3.** 11

Factorising and Solving Equations and Inequalities

Common Factors

You know how to multiply out brackets: $4(2x - 7) = 8x - 28$

The reverse process is called factorising: $8x - 28 = 4(2x - 7)$

Notice that $8x - 28 = 2(4x - 14)$ but the factor 2 is not the greatest factor common to 8 and 28 which is 4.

Always factorise by taking the greatest common factor outside the brackets

More examples:

$x^2 - 3x$	$24 - 8n$	$3a^2b - 6a$
$= x(x - 3)$	$= 8(3 - n)$	$= 3a(ab - 2)$

Top Tip
After factorising always check your answer by multiplying out the brackets.

Solving Equations – Cover-up Method

$2x + 1 = 9$ $3a - 1 = 2a + 5$ $m - 7 = 2$ $3(k + 2) = 9$

These are four examples of EQUATIONS. They all have the '=' sign

To 'solve' an equation means you must find the value the letter stands for that makes the equation true. Some equations can be solved by covering up the part that contains the letter:

Solve: $m - 7 = 2$

$- 7 = 2$

9 is covered if the equation is true

So $m = 9$

Solve: $2x + 1 = 9$

$+ 1 = 9$

8 is covered if the equation is true

So $2x = 8$

$2 \times$ $= 8$

4 is covered if the equation is true

So $x = 4$

Solve: $3(k + 2) = 9$

$3k + 6 = 9$

$+ 6 = 9$

3 is covered

So $3k = 3$

giving $k = 1$

Not all equations can be solved by covering up the part with the letter.

Solve: $3a - 1 = 2a + 5$

The letter appears on each side of the '=' sign

To solve this type of equation you need to use the balancing method.

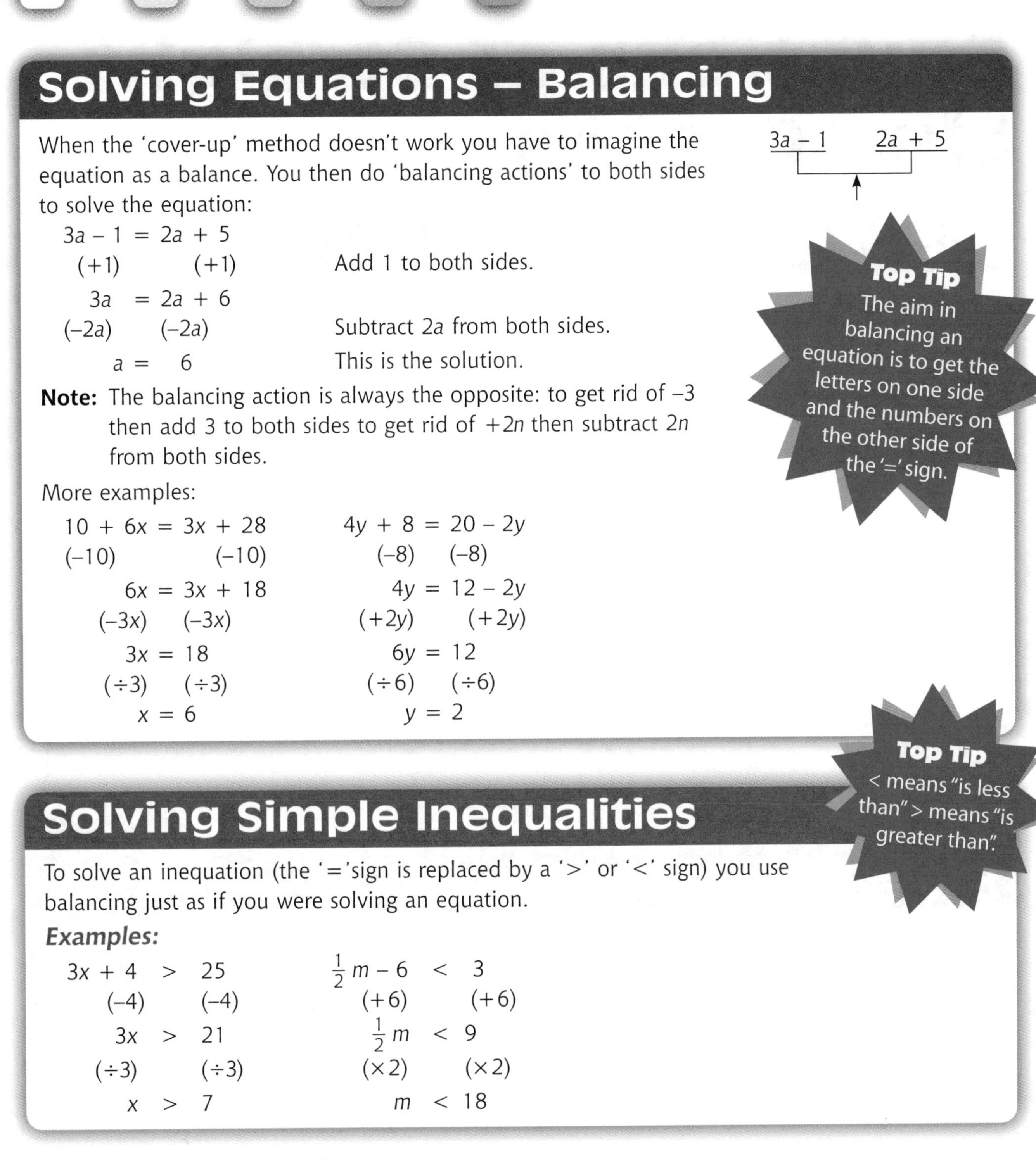

Solving Equations – Balancing

When the 'cover-up' method doesn't work you have to imagine the equation as a balance. You then do 'balancing actions' to both sides to solve the equation:

$3a - 1 = 2a + 5$

$(+1) \quad (+1)$ Add 1 to both sides.

$3a = 2a + 6$

$(-2a) \quad (-2a)$ Subtract $2a$ from both sides.

$a = 6$ This is the solution.

Note: The balancing action is always the opposite: to get rid of -3 then add 3 to both sides to get rid of $+2n$ then subtract $2n$ from both sides.

More examples:

$10 + 6x = 3x + 28$	$4y + 8 = 20 - 2y$
$(-10) \quad (-10)$	$(-8) \quad (-8)$
$6x = 3x + 18$	$4y = 12 - 2y$
$(-3x) \quad (-3x)$	$(+2y) \quad (+2y)$
$3x = 18$	$6y = 12$
$(\div 3) \quad (\div 3)$	$(\div 6) \quad (\div 6)$
$x = 6$	$y = 2$

Top Tip
The aim in balancing an equation is to get the letters on one side and the numbers on the other side of the '=' sign.

Solving Simple Inequalities

Top Tip
$<$ means "is less than" $>$ means "is greater than".

To solve an inequation (the '=' sign is replaced by a '$>$' or '$<$' sign) you use balancing just as if you were solving an equation.

Examples:

$3x + 4 > 25$	$\frac{1}{2}m - 6 < 3$
$(-4) \quad (-4)$	$(+6) \quad (+6)$
$3x > 21$	$\frac{1}{2}m < 9$
$(\div 3) \quad (\div 3)$	$(\times 2) \quad (\times 2)$
$x > 7$	$m < 18$

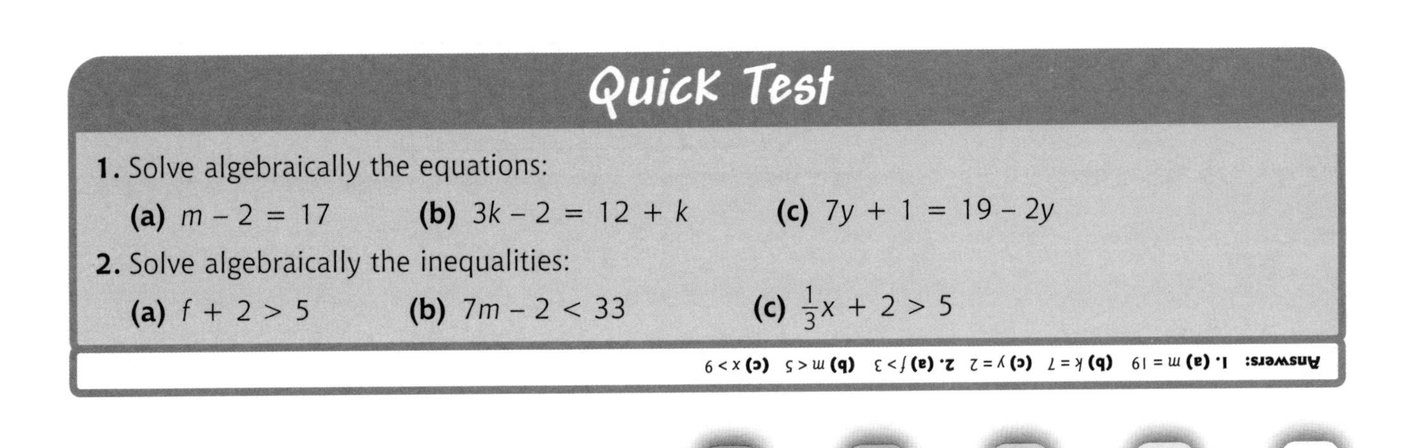

Quick Test

1. Solve algebraically the equations:

(a) $m - 2 = 17$ **(b)** $3k - 2 = 12 + k$ **(c)** $7y + 1 = 19 - 2y$

2. Solve algebraically the inequalities:

(a) $f + 2 > 5$ **(b)** $7m - 2 < 33$ **(c)** $\frac{1}{3}x + 2 > 5$

Answers: **1. (a)** $m = 19$ **(b)** $k = 7$ **(c)** $y = 2$ **2. (a)** $f > 3$ **(b)** $m < 5$ **(c)** $x > 9$

Equations of Straight Line Graphs

Tables of Values

You will be given a table with x values. Put each x value into the formula and complete the calculation to get the corresponding y-value. Enter these y values into the table.

Examples:

Formula: $y = 3x + 2$

Calculations

$x = 1 \quad y = 3 \times 1 + 2 = 5$
$x = 2 \quad y = 3 \times 2 + 2 = 8$
$x = 3 \quad y = 3 \times 3 + 2 = 11$

Table of Values

x	1	2	3
y	5	8	11

Formula: $y = \frac{1}{2}x - 1$

$x = -4 \quad y = \frac{1}{2} \times (-4) - 1 = -2 - 1 = -3$
$x = 0 \quad y = \frac{1}{2} \times 0 - 1 = 0 - 1 = -1$
$x = 4 \quad y = \frac{1}{2} \times 4 - 1 = 2 - 1 = 1$

x	−4	0	4
y	−3	−1	1

Formula: $y = -2x + 3$

$x = -2 \quad y = -2 \times (-2) + 3 = 4 + 3 = 7$
$x = 0 \quad y = -2 \times 0 + 3 = 0 + 3 = 3$
$x = 6 \quad y = -2 \times 6 + 3 = -12 + 3 = -9$

x	−2	0	6
y	7	3	−9

> **Top Tip**
> x is left/right (1st number)
> y is up/down (2nd number)

From a Table to a Graph

Table

x	0	1	2	3
y	2	4	6	8

(0, 2) (1, 4) (2, 6) (3, 8)

This table of values has four pairs of x and y values. You must change each pair into coordinates of a point.

Coordinates

(0, 2)
(1, 4)
(2, 6)
(3, 8)

In this case there are four points.

Plot these on a grid.

Draw a straight line graph through the points.

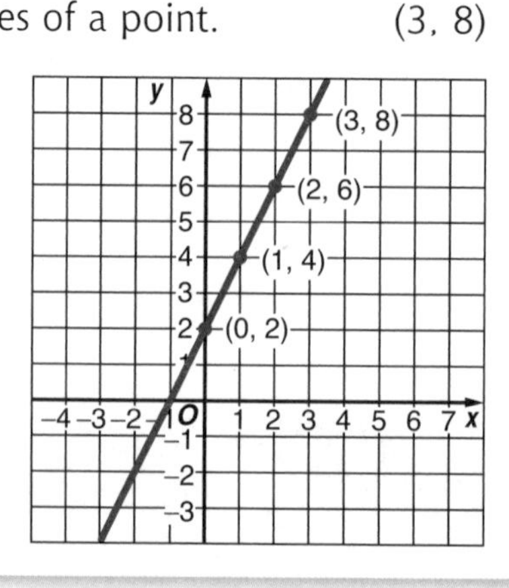

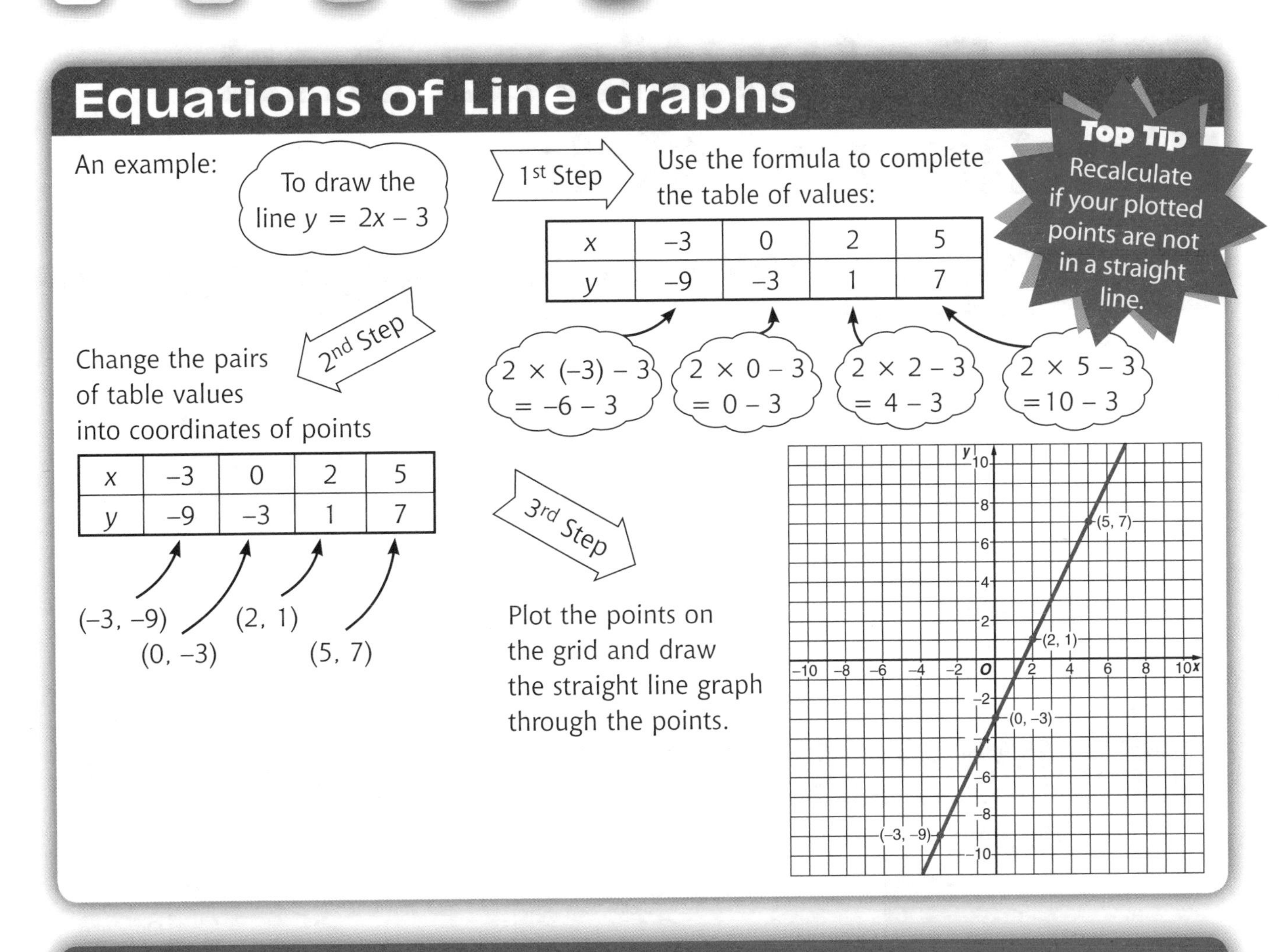

Special Lines

Examples:

$y = 2$

All points on the line $y = 2$ have a y-coordinate equal to 2.

(–3, 2) (0, 2) (1, 2)

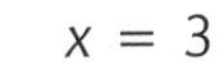

All points on the line $x = 3$ have a x-coordinate equal to 3.

(3, 4) (3, 1) (3, –1)

Quick Test

1. (a) Complete this table where $y = 2x + 1$

x	0	1	2	3
y				

(b) Use the table of values to draw the straight line $y = 2x + 1$ on a grid.

2. (a) Complete this table for $y = 0.5x - 1$

x	–4	0	6
y			

(b) Draw the line $y = 0.5x - 1$ on a grid.

3. Draw the lines

(a) $x = -3$

(b) $y = 1$ on a grid

4. Draw the line $y = 2 - x$ on a grid

Answers: 1. (a) 1, 3, 5, 7 **2. (a)** –3, –1, 2

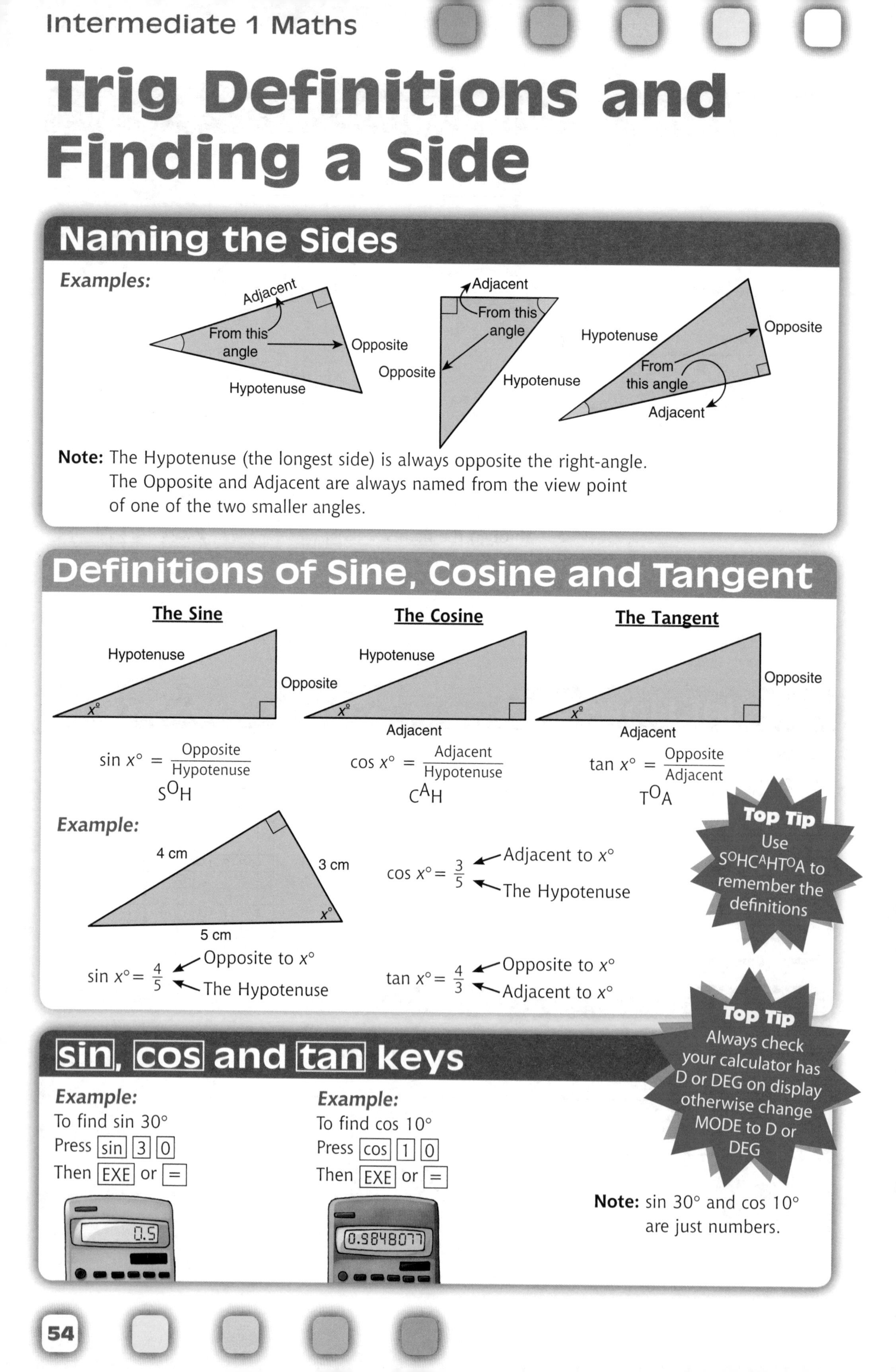

Trig Definitions and Finding a Side

Naming the Sides

Examples:

Note: The Hypotenuse (the longest side) is always opposite the right-angle. The Opposite and Adjacent are always named from the view point of one of the two smaller angles.

Definitions of Sine, Cosine and Tangent

The Sine

$\sin x° = \dfrac{\text{Opposite}}{\text{Hypotenuse}}$

S^{O}H

The Cosine

$\cos x° = \dfrac{\text{Adjacent}}{\text{Hypotenuse}}$

C^{A}H

The Tangent

$\tan x° = \dfrac{\text{Opposite}}{\text{Adjacent}}$

T^{O}A

Example:

$\cos x° = \frac{3}{5}$ ← Adjacent to $x°$ / ← The Hypotenuse

$\sin x° = \frac{4}{5}$ ← Opposite to $x°$ / ← The Hypotenuse

$\tan x° = \frac{4}{3}$ ← Opposite to $x°$ / ← Adjacent to $x°$

Top Tip
Use S^{O}HC^{A}HT^{O}A to remember the definitions

sin, cos and tan keys

Top Tip
Always check your calculator has D or DEG on display otherwise change MODE to D or DEG

Example:
To find sin 30°
Press [sin] [3] [0]
Then [EXE] or [=]

Example:
To find cos 10°
Press [cos] [1] [0]
Then [EXE] or [=]

Note: sin 30° and cos 10° are just numbers.

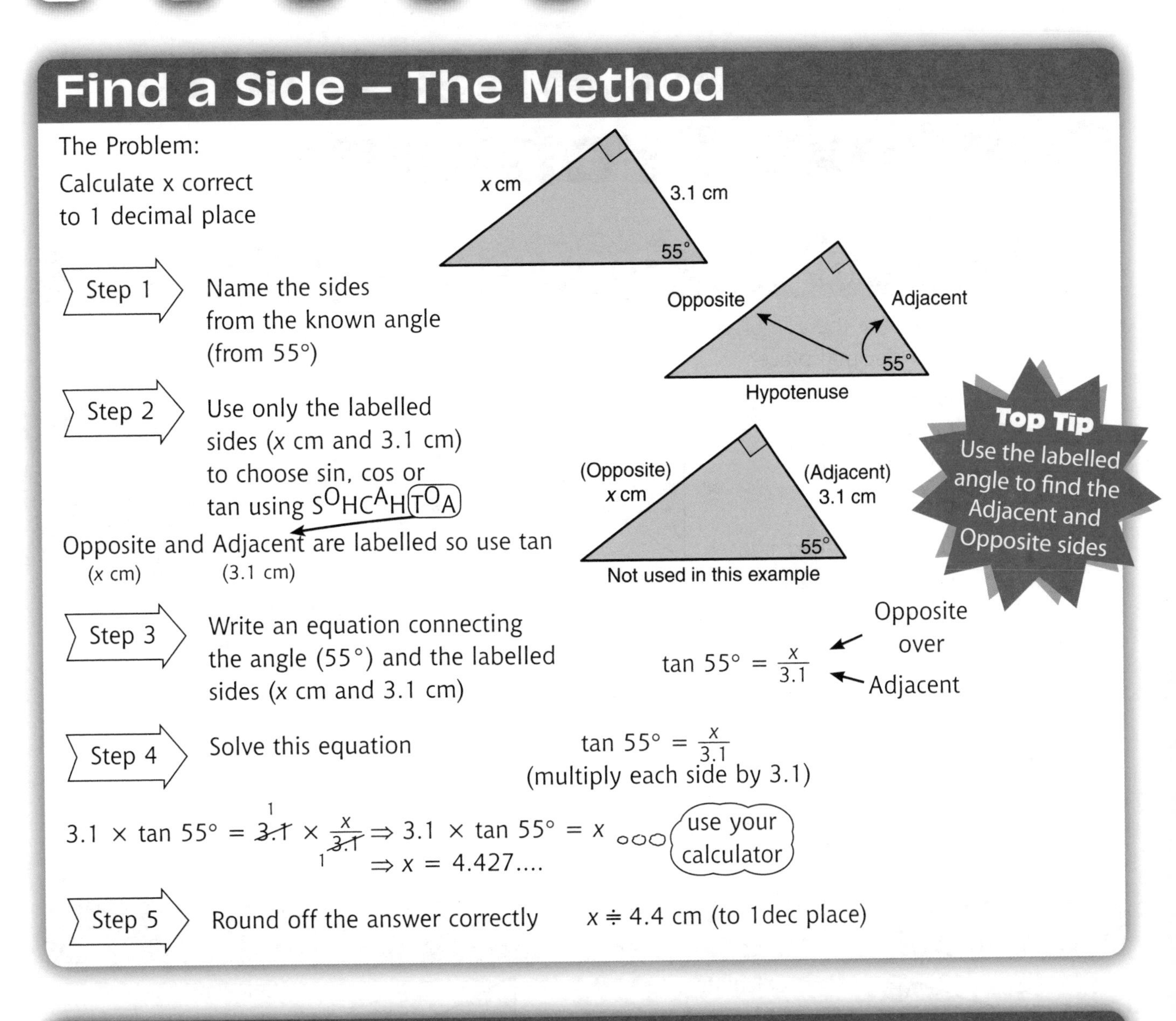

Find a Side – The Method

The Problem:
Calculate x correct to 1 decimal place

Step 1 Name the sides from the known angle (from 55°)

Step 2 Use only the labelled sides (x cm and 3.1 cm) to choose sin, cos or tan using $S^OH C^AH$ (T^OA)

Opposite and Adjacent are labelled so use tan
(x cm) (3.1 cm)

Top Tip
Use the labelled angle to find the Adjacent and Opposite sides

Step 3 Write an equation connecting the angle (55°) and the labelled sides (x cm and 3.1 cm)

$$\tan 55° = \frac{x}{3.1}$$

Opposite over Adjacent

Step 4 Solve this equation

$$\tan 55° = \frac{x}{3.1}$$

(multiply each side by 3.1)

$$3.1 \times \tan 55° = \overset{1}{\cancel{3.1}} \times \frac{x}{\underset{1}{\cancel{3.1}}} \Rightarrow 3.1 \times \tan 55° = x$$

$$\Rightarrow x = 4.427....$$

use your calculator

Step 5 Round off the answer correctly $x \doteqdot 4.4$ cm (to 1dec place)

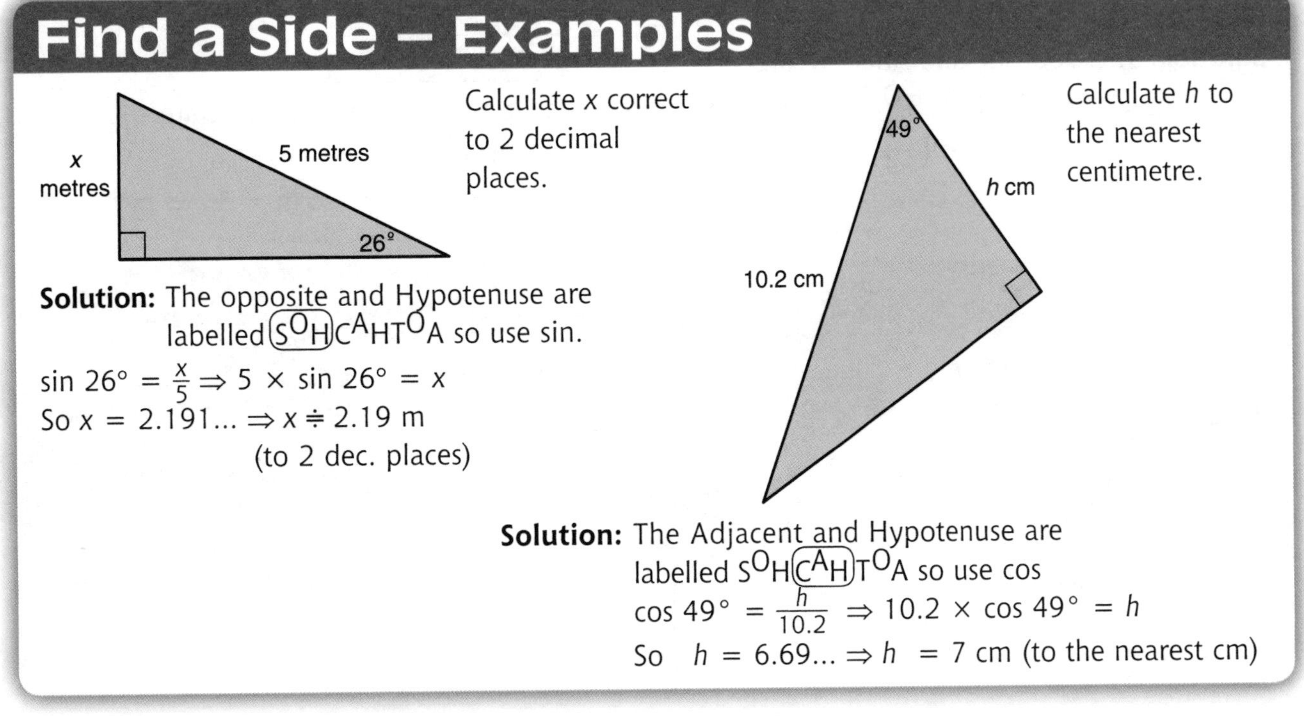

Find a Side – Examples

Calculate x correct to 2 decimal places.

Solution: The opposite and Hypotenuse are labelled (S^OH) $C^AH T^OA$ so use sin.

$\sin 26° = \frac{x}{5} \Rightarrow 5 \times \sin 26° = x$

So $x = 2.191... \Rightarrow x \doteqdot 2.19$ m
(to 2 dec. places)

Calculate h to the nearest centimetre.

Solution: The Adjacent and Hypotenuse are labelled S^OH (C^AH) T^OA so use cos

$\cos 49° = \frac{h}{10.2} \Rightarrow 10.2 \times \cos 49° = h$

So $h = 6.69... \Rightarrow h = 7$ cm (to the nearest cm)

Finding an Angle and Context Problems

Find an Angle – The Method

The Problem: Calculate the size of angle $x°$ correct to 1 decimal place

1.6 m, 2.2 m, $x°$

Step 1 Name the sides from the angle you are trying to find ($x°$)

Adjacent, Opposite, Hypotenuse, $x°$

Step 2 Use the labelled sides (1.6 m and 2.2 m) to choose sin, cos or tan from $S^OA\,C^AH\,T^OA$. In this case the Adjacent and the Hypotenuse are labelled. So use cos.

Step 3 Write an equation connecting the angle ($x°$) and the labelled sides (1.6 m and 2.2 m)

$$\cos x° = \frac{1.6}{2.2}$$ Adjacent over Hypotenuse

Step 4 Use cos^{-1} to find the angle

$$x = \cos^{-1}\left(\frac{1.6}{2.2}\right) = 43.341\ldots$$

Step 5 Round off the answer correctly $x° \doteqdot 43.3°$ (to 1 dec. place)

> **Top Tip**
> On some calculators you key [inv] [sin] for $\sin^{-1}$. Know your calculator!

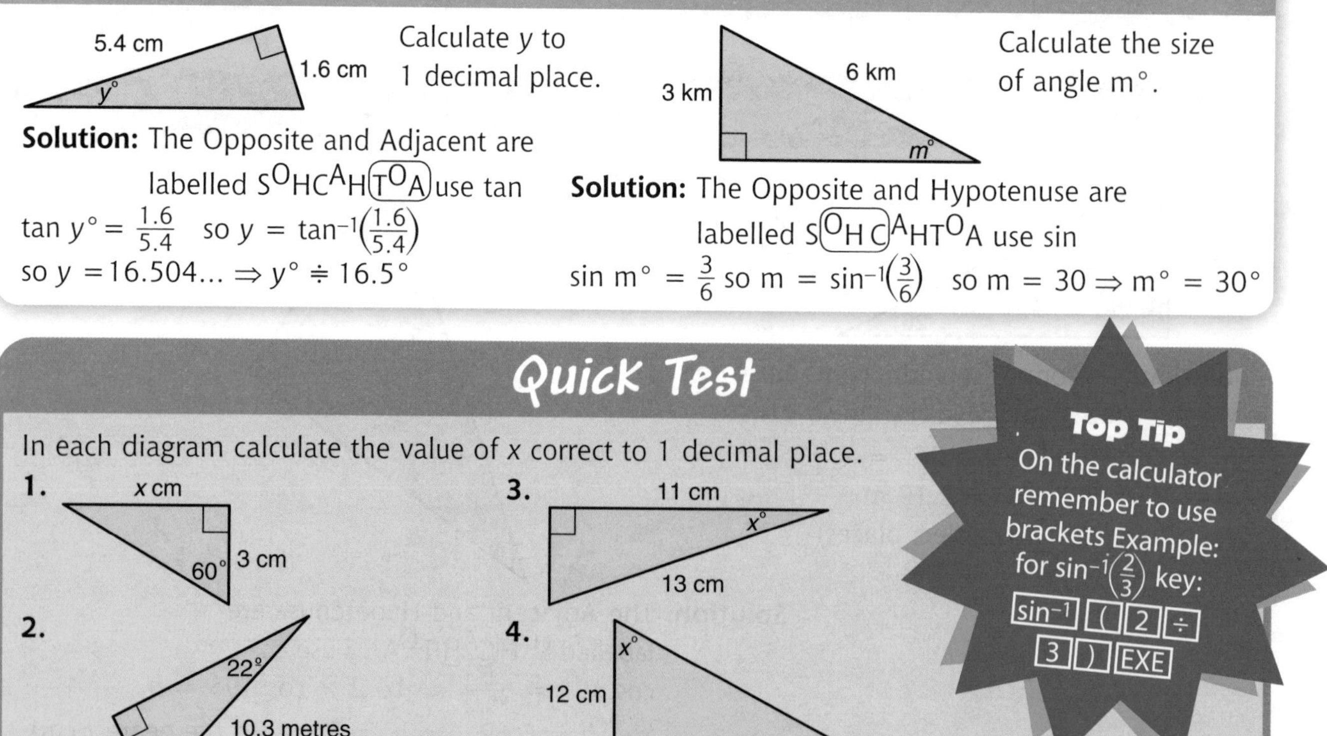

Find an Angle – Examples

5.4 cm, 1.6 cm, $y°$

Calculate y to 1 decimal place.

Solution: The Opposite and Adjacent are labelled $S^OH\,C^AH\,T^OA$ use tan

$\tan y° = \frac{1.6}{5.4}$ so $y = \tan^{-1}\left(\frac{1.6}{5.4}\right)$

so $y = 16.504\ldots \Rightarrow y° \doteqdot 16.5°$

3 km, 6 km, $m°$

Calculate the size of angle $m°$.

Solution: The Opposite and Hypotenuse are labelled $S^OH\,C^AH\,T^OA$ use sin

$\sin m° = \frac{3}{6}$ so $m = \sin^{-1}\left(\frac{3}{6}\right)$ so $m = 30 \Rightarrow m° = 30°$

Quick Test

In each diagram calculate the value of x correct to 1 decimal place.

1. x cm, 3 cm, 60°
2. 22°, 10.3 metres, x metres
3. 11 cm, 13 cm, $x°$
4. $x°$, 12 cm, 23 cm

> **Top Tip**
> On the calculator remember to use brackets Example: for $\sin^{-1}\left(\frac{2}{3}\right)$ key: [sin^{-1}] [(] [2] [÷] [3] [)] [EXE]

Answers: 1. 5.2 cm 2. 3.9 metres 3. 32.2 4. 62.4

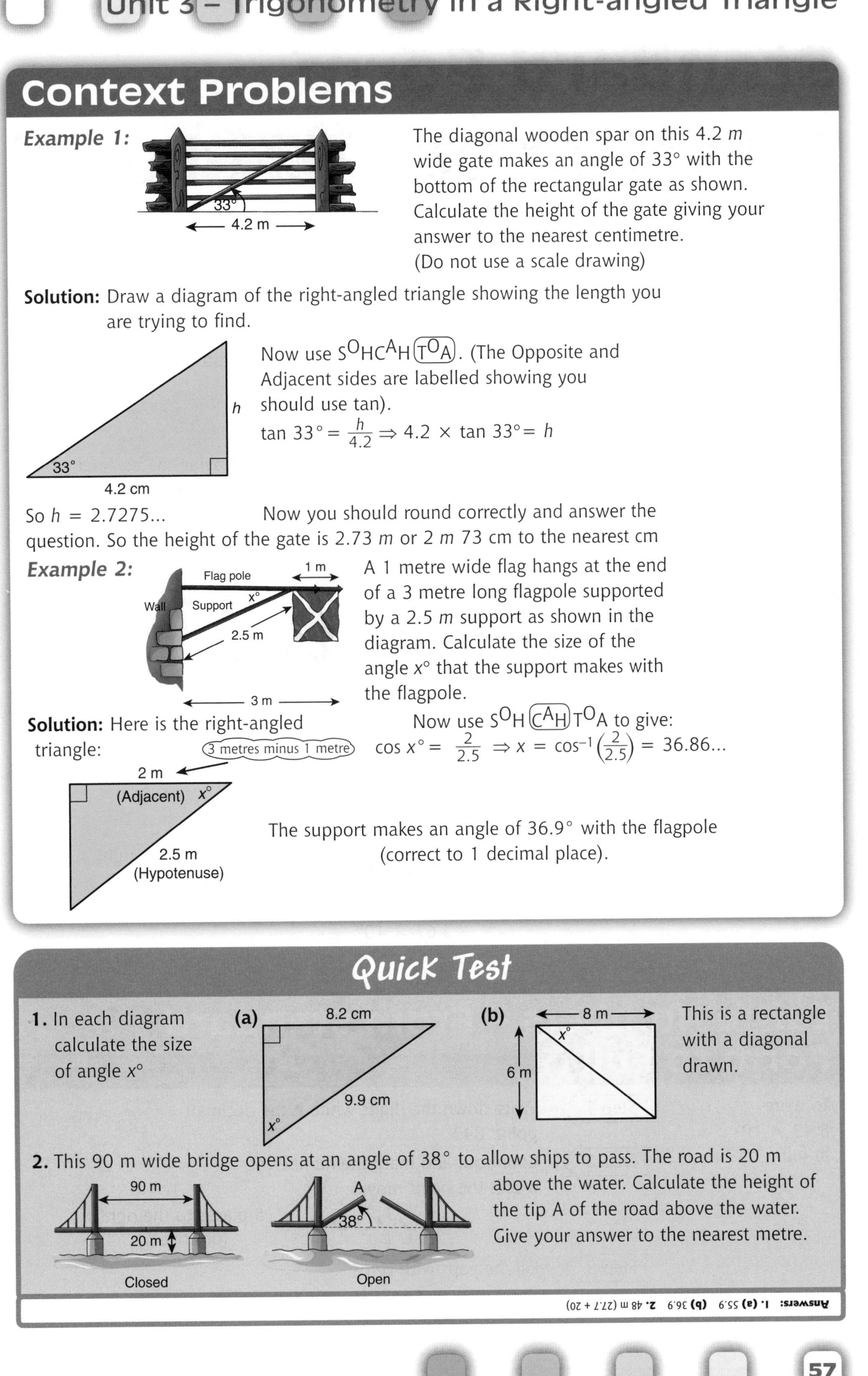

Context Problems

Example 1:

The diagonal wooden spar on this 4.2 *m* wide gate makes an angle of 33° with the bottom of the rectangular gate as shown. Calculate the height of the gate giving your answer to the nearest centimetre. (Do not use a scale drawing)

Solution: Draw a diagram of the right-angled triangle showing the length you are trying to find.

Now use S^{O}HC^{A}H(T^{O}A). (The Opposite and Adjacent sides are labelled showing you should use tan).

$\tan 33° = \frac{h}{4.2} \Rightarrow 4.2 \times \tan 33° = h$

So $h = 2.7275...$ Now you should round correctly and answer the question. So the height of the gate is 2.73 *m* or 2 *m* 73 cm to the nearest cm

Example 2:

A 1 metre wide flag hangs at the end of a 3 metre long flagpole supported by a 2.5 *m* support as shown in the diagram. Calculate the size of the angle $x°$ that the support makes with the flagpole.

Solution: Here is the right-angled triangle:

Now use S^{O}H(C^{A}H)T^{O}A to give:

$\cos x° = \frac{2}{2.5} \Rightarrow x = \cos^{-1}\left(\frac{2}{2.5}\right) = 36.86...$

The support makes an angle of 36.9° with the flagpole (correct to 1 decimal place).

Quick Test

1. In each diagram calculate the size of angle $x°$

(a)

(b) This is a rectangle with a diagonal drawn.

2. This 90 m wide bridge opens at an angle of 38° to allow ships to pass. The road is 20 m above the water. Calculate the height of the tip A of the road above the water. Give your answer to the nearest metre.

Answers: **1. (a)** 55.9 **(b)** 36.9 **2.** 48 m (27.7 + 20)

Standard Form

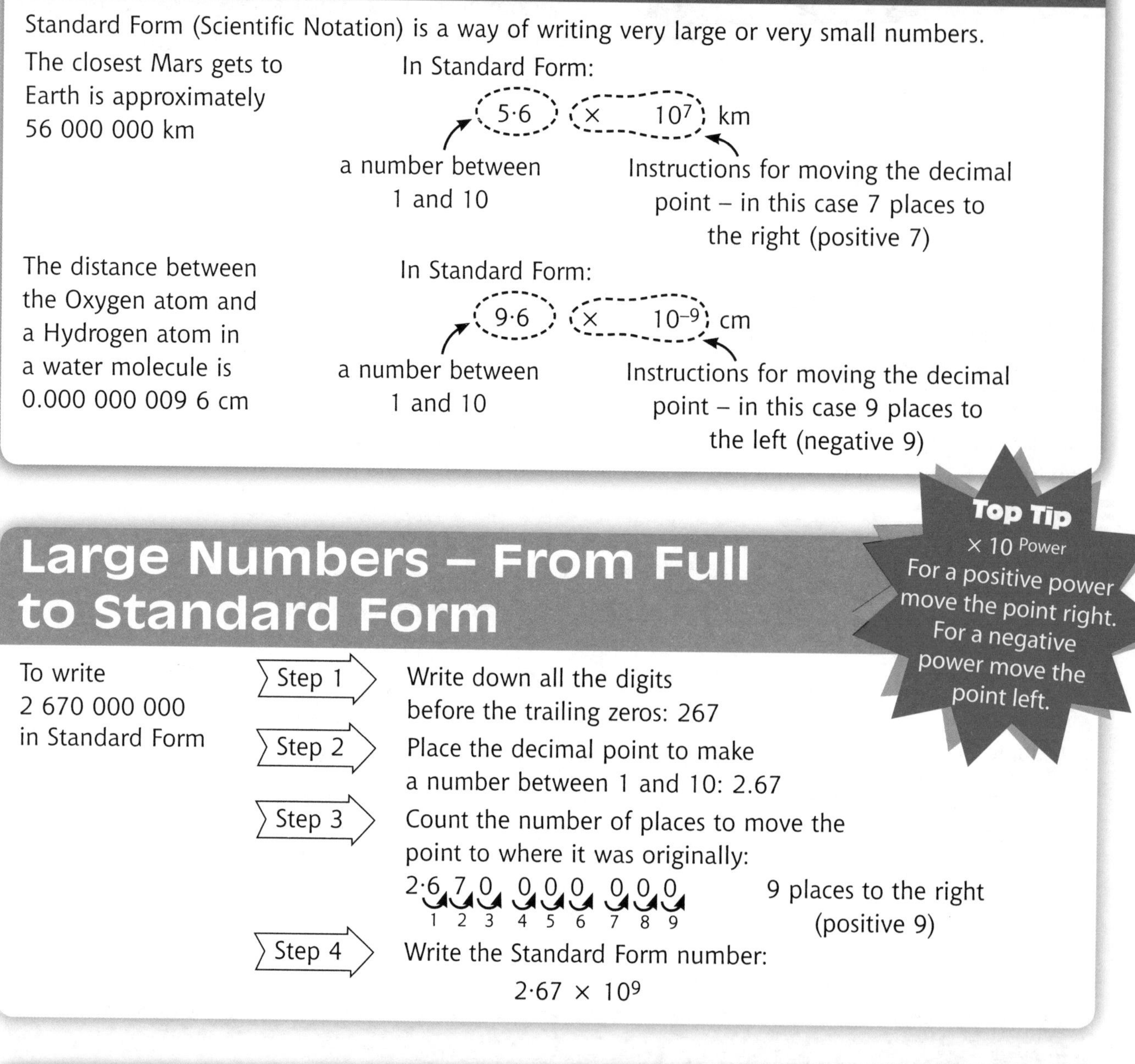

What is Standard Form?

Standard Form (Scientific Notation) is a way of writing very large or very small numbers.

The closest Mars gets to Earth is approximately 56 000 000 km

In Standard Form:

$5{\cdot}6 \times 10^7$ km

a number between 1 and 10 (5·6)

Instructions for moving the decimal point – in this case 7 places to the right (positive 7) ($\times 10^7$)

The distance between the Oxygen atom and a Hydrogen atom in a water molecule is 0.000 000 009 6 cm

In Standard Form:

$9{\cdot}6 \times 10^{-9}$ cm

a number between 1 and 10 (9·6)

Instructions for moving the decimal point – in this case 9 places to the left (negative 9) ($\times 10^{-9}$)

Top Tip

$\times 10$ Power

For a positive power move the point right. For a negative power move the point left.

Large Numbers – From Full to Standard Form

To write 2 670 000 000 in Standard Form

Step 1 Write down all the digits before the trailing zeros: 267

Step 2 Place the decimal point to make a number between 1 and 10: 2.67

Step 3 Count the number of places to move the point to where it was originally:

2·6 7 0 0 0 0 0 0 0 (moves 1 2 3 4 5 6 7 8 9) — 9 places to the right (positive 9)

Step 4 Write the Standard Form number:

$2{\cdot}67 \times 10^9$

Large Numbers – From Standard Form to Full

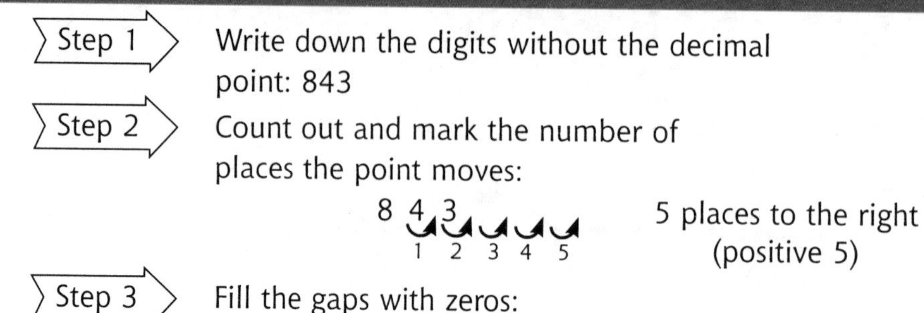

To write 8.43×10^5 in full.

Step 1 Write down the digits without the decimal point: 843

Step 2 Count out and mark the number of places the point moves:

8 4 3 _ _ _ (moves 1 2 3 4 5) — 5 places to the right (positive 5)

Step 3 Fill the gaps with zeros:

8 4 3 0 0 0

Small Numbers – From Full to Standard Form

To write 0.000 008 4 in Standard Form

Step 1 Write down all the digits after the leading zeros: 84

Step 2 Place the decimal point to give a number between 1 and 10: 8.4

Step 3 Count the number of places to move the point to where it was originally:

0 0 0 0 0 0 8 4

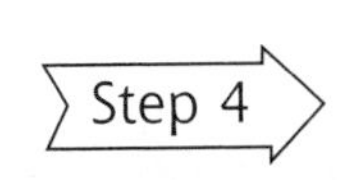

6 places to the left (negative 6)

Step 4 Write the number in Standard Form:

8.4×10^{-6}

Small Numbers – From Standard Form to Full

To write $2{\cdot}68 \times 10^{-7}$ in full.

Step 1 Write down the digits without the decimal point:

2 6 8

Step 2 Count out and mark the number of places the point moves:

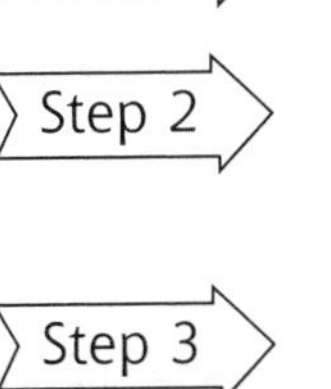

2 6 8

7 places to the left (negative 7)

Step 3 Fill the gaps with zeros and add an extra zero in front of the decimal point. 0.000 000 268

Using your Calculator for Standard Form

The [EXP] button is used to enter Standard Form numbers into your calculator:

Examples: 2.36×10^{4} enter: [2] [·] [3] [6] [EXP] [4]

8.4×10^{-5} enter: [8] [·] [4] [EXP] [(–)] [5] ← [(–)] is the 'negative' key

Context Example:

A rod-shaped virus has length 1.3×10^{-6} cm. Calculate the length of 50 000 of these viruses lying end-to-end. Give your answer in Standard Form.

Solution: $50\,000 \times (1.3 \times 10^{-6})$ ← The calculation is [5][0][0][0][0][×][1][·][3][EXP][(–)][6]

$= 0.065$ cm

$= 6.5 \times 10^{-2}$ cm

Top Tip
The [EXP] key stands for "times ten to the power of"

Quick Test

1. Write these numbers in full: **(a)** 2.8×10^{4} **(b)** 3.9×10^{-3}
2. Write these numbers in Standard Form: **(a)** 79 300 000 **(b)** 0.000 029
3. Light travels approximately 1.8×10^{7} km in one minute. How far does it travel in one second? Write your answer in Standard Form.

Answers: 1. (a) 28000 **(b)** 0.0039 **2. (a)** 7.93×10^{7} **(b)** 2.9×10^{-5} **3.** 3×10^{5} km.

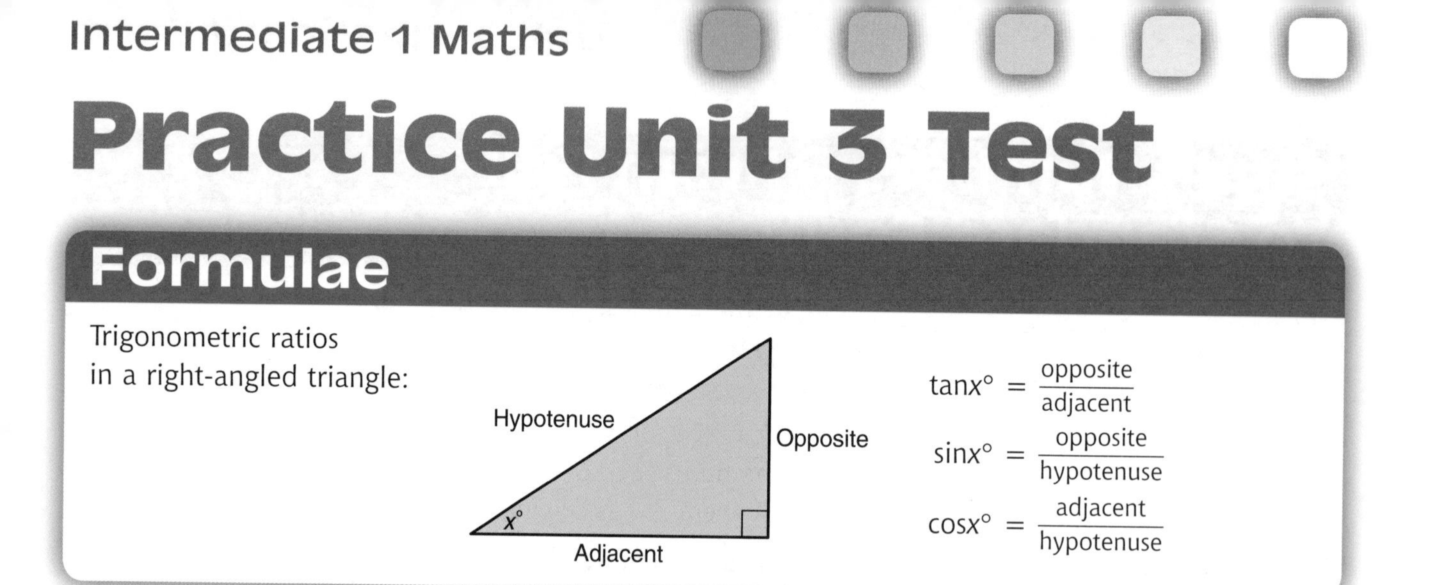

Practice Unit 3 Test

Formulae

Trigonometric ratios in a right-angled triangle:

$$\tan x° = \frac{\text{opposite}}{\text{adjacent}}$$

$$\sin x° = \frac{\text{opposite}}{\text{hypotenuse}}$$

$$\cos x° = \frac{\text{adjacent}}{\text{hypotenuse}}$$

Outcome 1

1. Given that $A = p - qr$ calculate A when $p = 17$, $q = 3$ and $r = 4$

2. (a) Multiply out the brackets $5(m - 2)$

(b) Simplify this expression $3(y - 1) + 2y$

3. Factorise $6x + 18$

4. Solve the equation: **(a)** $n - 8 = 11$ **(b)** $6y = 8$

5. Solve the inequality:

(a) $k + 1 > 8$ **(b)** $5m < 30$

Outcome 2

6. (a) Complete this table where $y = 2x - 1$

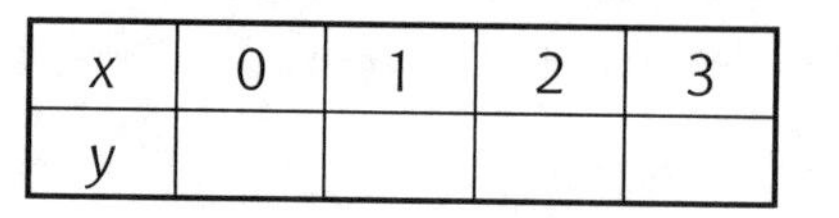

x	0	1	2	3
y				

(b) Use the table of values to draw the straight line $y = 2x - 1$ on the grid.

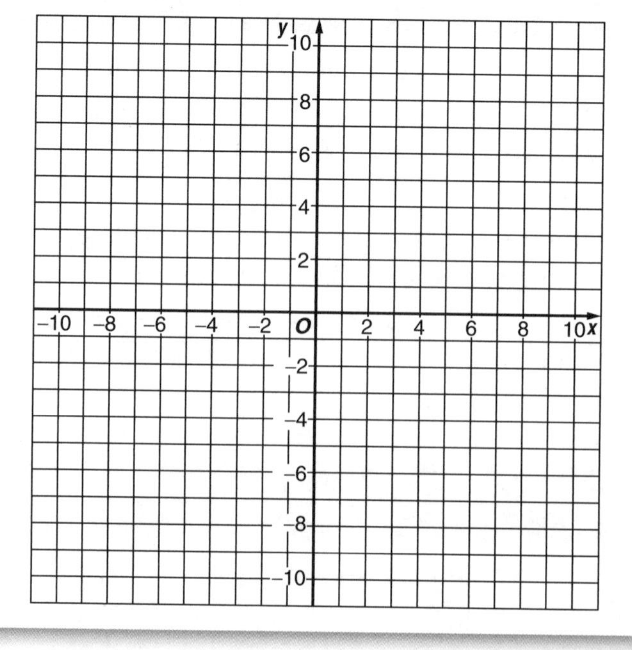

Outcome 3

7. 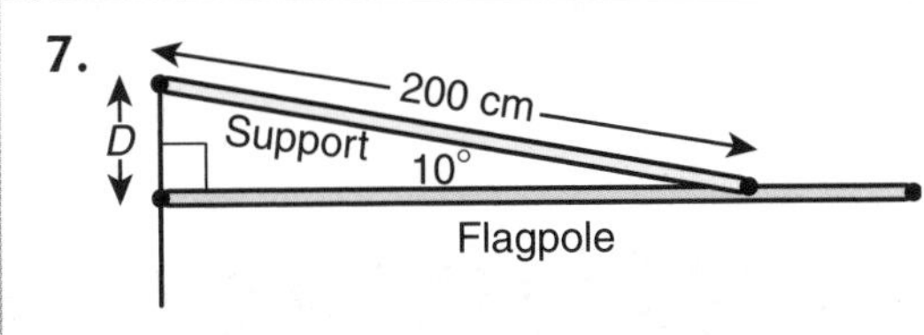

The diagram shows a 200 cm long support holding up a flagpole. The support makes an angle of 10° with the flagpole. Calculate D, the distance from the end of the support to the end of the flagpole.

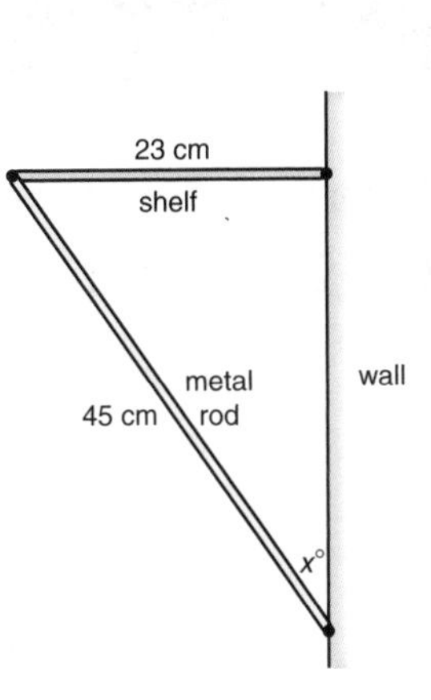

8. The diagram shows a horizontal shelf attached to a vertical wall supported by a 45 cm metal rod. The shelf is 23 cm wide and the rod makes an angle of $x°$ with the wall as shown in the diagram. Calculate the size of the angle $x°$.

Outcome 4

9. **(a)** An astronomer estimated that our Milky Way Galaxy contains around 3.8×10^{11} stars. Write this number out in full.

   (b) The common cold virus is approximately 2.1×10^{-5} cm in diameter. Write this number out in full.

10. **(a)** In 2009 the world population was estimated at 6 900 000 000. Write this number in standard form.

   (b) A spider spins a silk thread that is roughly 0.00015 cm thick. Write this number in standard form.

11. Sound travels approximately 3.43×10^4 cm in one second. How far does it travel in 1 minutes (60 seconds)? Write your answer in Standard Form.

Preparation for Assessment

Practice Unit 3 Questions

1. Solve algebraically the equation $8x - 7 = 21 + x$

2. A biologist found the average size of a clover leaf cell to be 2.86×10^{-3} cm. Write this number out in full.

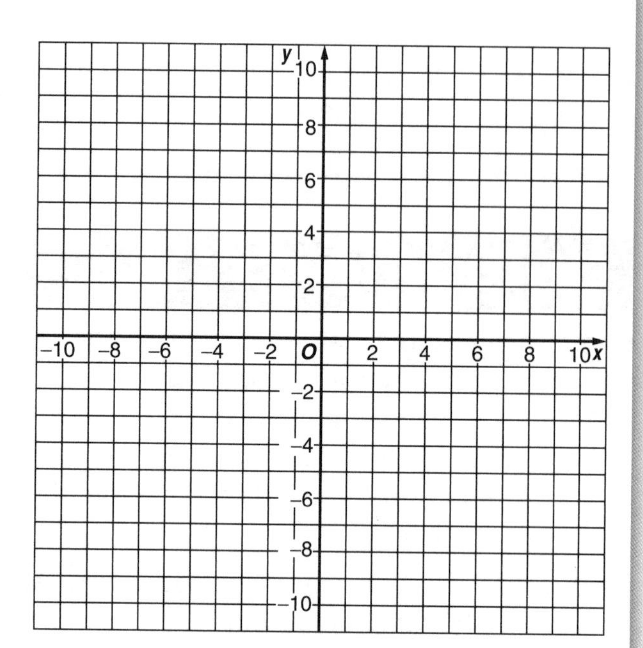

3. (a) Complete the table below for $y = 1.5x - 2$

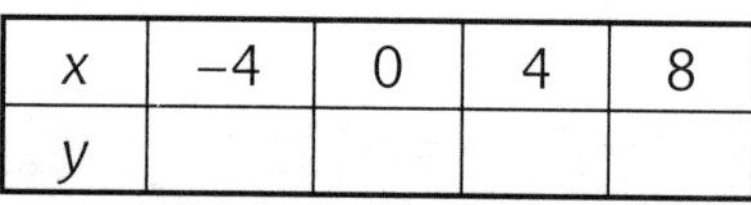

x	-4	0	4	8
y				

(b) Draw the two lines on the grid.

(i) $y = 1.5x - 2$ **(ii)** $x = -4$

4. Evaluate $b - a^2$ when $a = -3$ and $b = 8$

5. (a) Multiply out the brackets and simplify $3(6m - 1) + 7$

(b) Factorise $8x + 12$

6. Solve algebraically the inequality $\frac{1}{3}x - 1 < 2$

7. Calculate the area of the right-angled triangle shown in the diagram. Do not use a scale drawing.

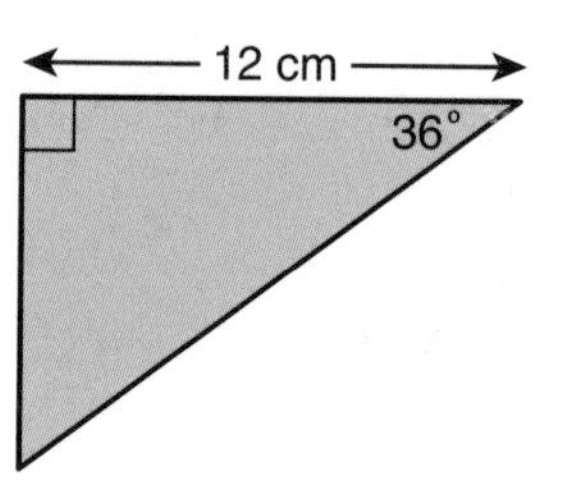

8. Use the formula $C = \frac{L(a - b)}{a}$ to find the value of C when $L = 12.8$, $a = 6.4$ and $b = 0.4$

9. Solve algebraically the inequality $6x + 4 > 58$

10. Theformula for the semiperimeter, s cm, of a triangle is

$$s = \frac{1}{2}(a + b + c)$$

where a, b and c are the lengths of the three sides of the triangle. Calculate s for the triangle shown.

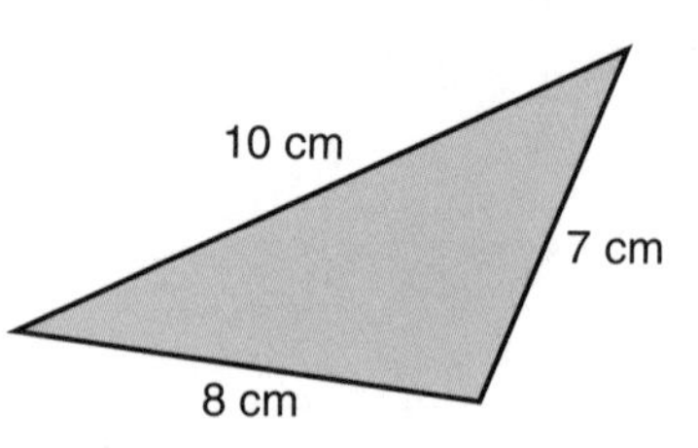

11. (a) Complete the table below for $y = \frac{1}{2}x + 1$

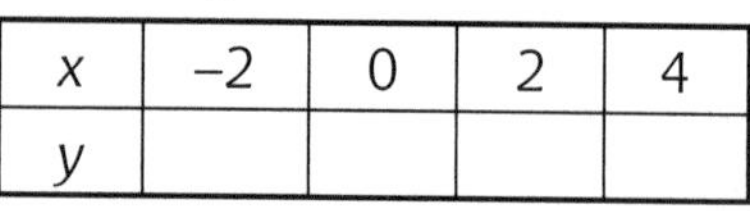

x	-2	0	2	4
y				

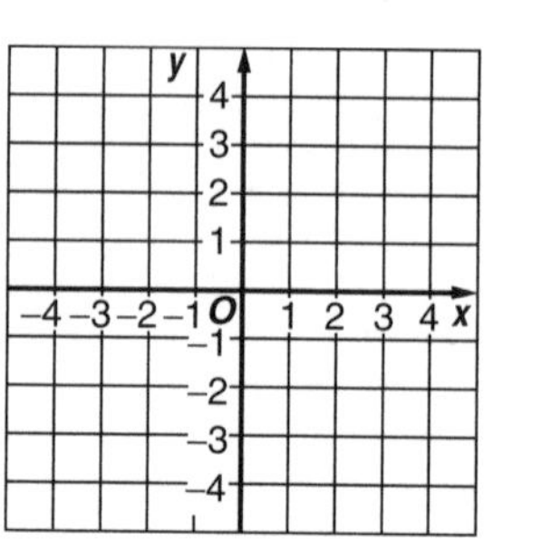

(b) Draw the line

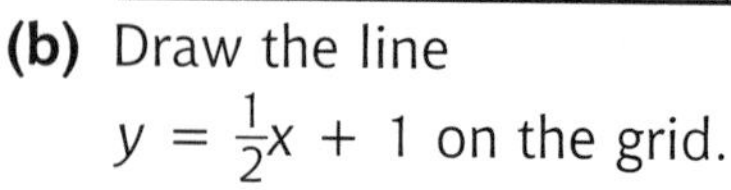

$y = \frac{1}{2}x + 1$ on the grid.

12. In1974 a fossil skeleton of a female ancestor of humans was discovered and was given the name Lucy. It is estimated that Lucy lived 3.2 million years ago. Write this number in Standard Form.

13. Solve algebraically the equation $13n - 8 = 2n + 113$

14. **(a)** Multiply out the brackets and simplify $2(6x + 1) - 4x$

(b) Factorise $24 - 20n$

15. Thediagram shows the end of a garden hut in the shape of a rectangle and a right-angled triangle. The sloping roof makes an angle of 25° with the horizontal. Calculate the height of the point A above the ground. Give your answer correct to 1 decimal place.

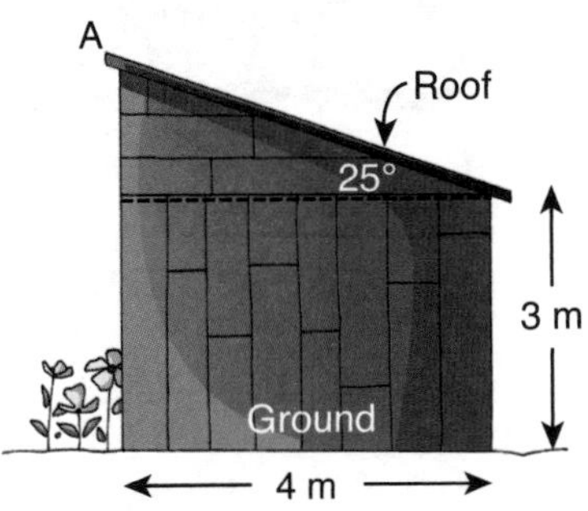

16. Usethe formula $L = \frac{1}{2}ab^2$ to find the value of L when $a = 2$ and $b = 4$.

17. Lighttravels one mile in about 5.4×10^{-6} seconds. Calculate how long light takes to travel 5000 miles. Write your answer in Standard Form.

18. **(a)** Multiply out the brackets and simplify $5k + 3(m - 4k)$

(b) Factorise $16n - 24$

Pay – Basic Terms and Calculation of Gross Pay

Reminders from Unit 1

There are various ways you can get paid for the work you do. Here are some of the terms you should know:

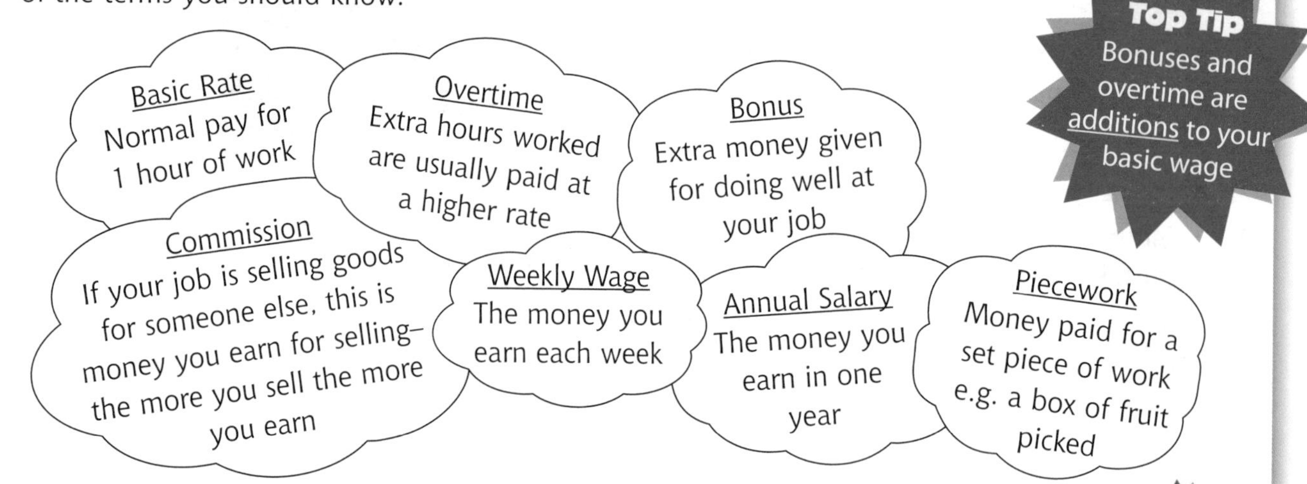

Top Tip
Bonuses and overtime are additions to your basic wage

Gross Pay, Deductions and Net Pay

Top Tip
Net Pay equals Gross Pay minus Deductions

Gross Pay is the money you earn before deductions are made

Net Pay is the money you are left with after deductions are taken off.

So part of the money you earn (your earnings) will not be paid to you. The sums of money that are taken off are called deductions.

The most common Deductions are:

Income Tax: This is used to help the government pay for running the country – education, defence etc

National Insurance: This is used to pay for the National Health Service, pensions (money paid to retired people) and unemployment benefit (money paid to unemployed people)

Other deductions may be for superannuation (extra pension payments) or payments to a Union or for extra insurance etc.

Calculation of Gross Pay

Gross Pay = Basic Pay + Bonuses and Overtime

Example: Pete is a plumber and earns a weekly basic wage of £250. Calculate his Gross Pay in a week where he earns £85 overtime and a bonus of £25.

Solution: Gross Pay = Basic Pay + Overtime + Bonus
= £250 + £85 + £25
= £360

Top Tip
Overtime and bonuses are ADDITIONS to your pay. Income Tax and Insurance are SUBTRACTED from your pay.

Payslips

When you get your pay you will also be given a payslip showing.

- Your gross pay – basic pay plus any additions (bonuses, overtime, etc) before any deductions are taken off
- Your deductions – income tax, national insurance, etc
- Your net pay – what you take home after any deductions are taken off

Mr. S. Leckie	
Payments	
Basic Pay	3,543.40
Total Payments	3,543.40
Deductions	
Income Tax	728.20
National Insurance	268.73
Total Deductions	996.93
Net Pay:	**2,546.47**

Example:

Part of Jack's payslip is shown. Complete his payslip to show the net pay.

Solution: Net Pay = Gross Pay – Deductions
= £525.40 – £201.80
= £323.60

Employee Name: Jack M Cooper	
Payments	
Gross Pay:	525.40
Deductions	
Total Deductions:	201.80
Net Pay:	

Rates of Pay and Overtime Payments

Calculating Gross Pay

Your basic rate of pay is how much you are paid for 1 hours work

Example:

Work out Ethan's Gross weekly pay if he works 35 hours at £7.50 per hour

Solution: Gross Pay = 35 × £7.50
= £262.50

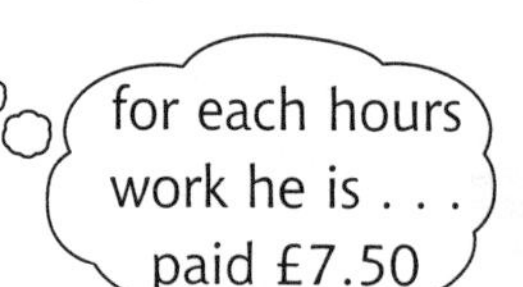

Advert
EARN £6.50 per hour delivering leaflets

Gardener Wanted
£12 per hour offered to a hardworking handyman

Brick layers wanted
Excellent rates of pay offered for ...

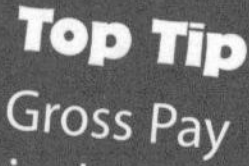

Top Tip
Gross Pay includes bonuses and overtime

Gross Pay on a Payslip

Your payslip will show your basic pay and any bonuses and overtime payments.

Example:

Complete Evie's weekly payslip if she worked 38 hours at £12.20 per hour during the week. She received a bonus of £10 and her overtime for the week was £24.40.

Employee Name: Evie Crichton

Payments

Basic Pay:	£ ______
Bonuses:	£ ______
Overtime:	£ ______
Total Gross Pay:	£ ______

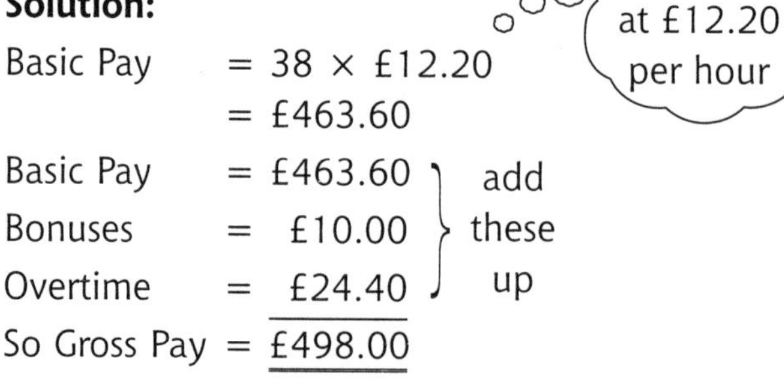

Solution:

Basic Pay = 38 × £12.20
= £463.60

38 hours at £12.20 per hour

Basic Pay = £463.60
Bonuses = £10.00
Overtime = £24.40
(add these up)
So Gross Pay = £498.00

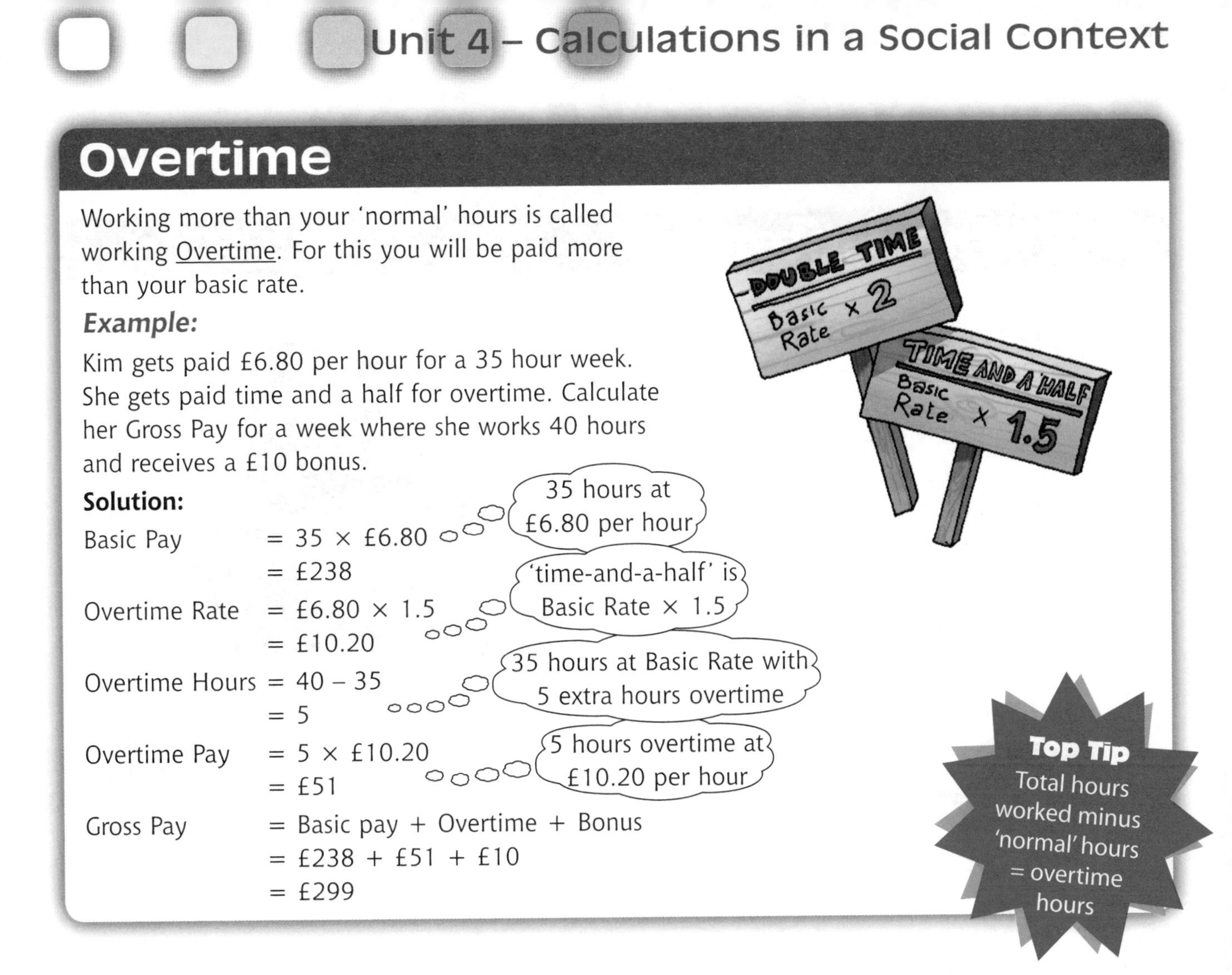

Overtime

Working more than your 'normal' hours is called working <u>Overtime</u>. For this you will be paid more than your basic rate.

Example:

Kim gets paid £6.80 per hour for a 35 hour week. She gets paid time and a half for overtime. Calculate her Gross Pay for a week where she works 40 hours and receives a £10 bonus.

Solution:

Basic Pay = 35 × £6.80
= £238

(35 hours at £6.80 per hour)

Overtime Rate = £6.80 × 1.5
= £10.20

('time-and-a-half' is Basic Rate × 1.5)

Overtime Hours = 40 – 35
= 5

(35 hours at Basic Rate with 5 extra hours overtime)

Overtime Pay = 5 × £10.20
= £51

(5 hours overtime at £10.20 per hour)

Gross Pay = Basic pay + Overtime + Bonus
= £238 + £51 + £10
= £299

Top Tip
Total hours worked minus 'normal' hours = overtime hours

Quick Test

1. Megan's basic rate of pay is £8.80 per hour. How much is she paid for 3 hours overtime at wtime and a half?
2. Olivia works in an office for a basic rate of £6.40 per hour. Any overtime she does is paid at double time. Complete her payslip for a week in which she works 4 hours overtime.

Employee Name: Olivia Newton

Payments

	Hours	Rate	Amount
Basic Pay:	32	£6.40	
Overtime:	4		
Total Gross Pay:			

Answers: 1. £39.60 **2.**

	Hours	Rate	Amount
Basic Pay	32	£6.40	£204.80
	4	£12.80	£51.20
		Total Gross	£256.00

Borrowing Money

Loans and Interest

People borrow money when they don't have enough cash to pay for what they want to buy. This money is called a LOAN.

You will eventually have to pay back the loan and will also be asked to pay extra money. This is called INTEREST and is the cost of the loan.

When you take out a loan you should be told:

- The monthly Rate of Interest
- The Annual Percentage Rate (APR)

Where can you borrow money

Banks

Building Societies

Credit Companies

Shops

Example: The loan is charged at a monthly rate of interest of 1.5% with an APR of 20%

Calculate **(a)** The interest after one month
(b) The amount owed after one month
(c) The interest due after one year.

A £1200 loan.

Top Tip
To find 1% divide by 100

Solution: (a) 1.5% of £1200 = $\frac{1.5}{100} \times 1200 = £18$
The interest for one month is £18.

(b) The amount owed after one month is
£1200 + £18 = £1218 (loan + interest)

(c) 20% of £1200 = $\frac{20}{100} \times 1200 = £240$
The interest for one year is £240

Top Tip
The APR rate gives interest for a year

Loan Protection

You can choose to pay extra each month when you take out a loan. This is for Loan Protection. This means if a misfortune happens (you lose your job, you have an accident etc) your loan repayments will be paid for you for the length of time that you cannot afford to pay them.

Loan Tables

Here is a loan table:

Monthly payments without Loan protection

LOAN PERIOD	AMOUNT		
	£500	£1000	£2000
6 months	£86.30	£172.60	£345.20
12 months	£44.43	£88.85	£177.70
18 months	£30.50	£60.99	£121.98
24 months	£23.54	£47.08	£94.15

Monthly payments with Loan protection

LOAN PERIOD	AMOUNT		
	£500	£1000	£2000
6 months	£93.20	£186.41	£372.82
12 months	£47.98	£95.96	£191.92
18 months	£32.94	£65.87	£131.74
24 months	£25.42	£50.85	£101.68

Example:

Mr Cooper takes out a loan of £1000 without Loan Protection and decides to pay it back over the course of 18 months

Use the left hand table for no Loan Protection

	£1000
18 months	£60.99

The table shows he will pay £60.99 each month for the 18 months of the loan $18 \times £60.99 = £1097.82$

This is the repayment total. Notice the Loan has cost him £97.82 in interest.

Mrs Gillespie borrows £500 with Loan Protection and pays back the loan over 6 months.

Use the right hand table for Loan Protection

	£500
6 months	£93.20

The table shows she will pay £93.20 each month for the 6 months of the loan $6 \times £93.20 = £559.20$

Her interest is £59.20 with the total repayment being £559.20

Top Tip
Loan tables show Monthly repayments

Quick Test

1. (a) Mr Fordyce took out a £2000 loan with Loan Protection over 24 months. Use the table above to find:

(i) his monthly repayment

(ii) The total repayments after 24 months

(b) After 18 months Mr Fordyce became ill and the loan company paid the remaining monthly repayments. How much did the company pay?

Answers: 1. (a)(i) £101.68 (ii) £2440.32 (b) £610.08

Network Diagrams

What is a Network Diagram?

Here is a map showing part of Fife.

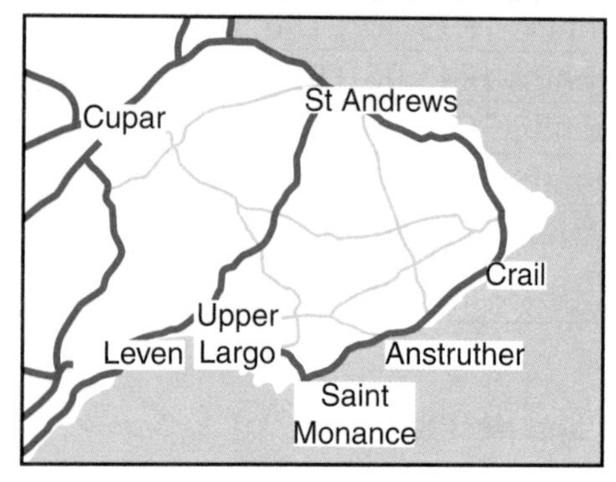

A Network Diagram can be used to show how the towns are connected by main roads:

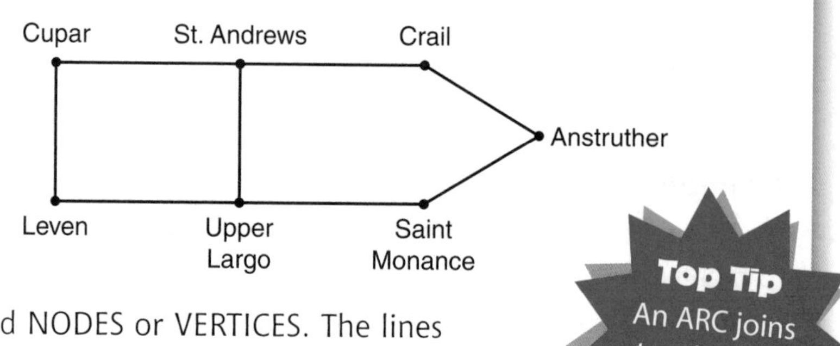

Top Tip
An ARC joins two NODES together.

The dots in a Network Diagram are called NODES or VERTICES. The lines joining the nodes are called ARCS.

Here is the same diagram with the distances between the towns shown.

Notice that a Network Diagram is not to scale and only shows connections.

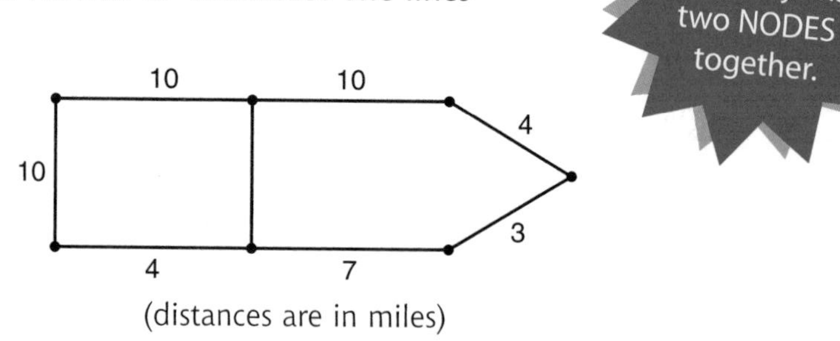

(distances are in miles)

Order of a Node

arc arc arc node Upper Largo

3 arcs meet at the Upper Largo node.

This node has order 3

arc arc Anstruther node

2 arcs meet at the Anstruther node.

This node has order 2

Top Tip
Count all the lines out of a node to get the ORDER

Example:

B C D A F E

Which nodes have order 4 in this network diagram?

Solution: Each of the nodes B,C, E and F have 4 arcs meeting at them and so have order 4.

Using Network Diagrams

Example:

This network diagram shows the time it takes two people to build a set of drawers from a 'flat pack'. All times are in minutes.

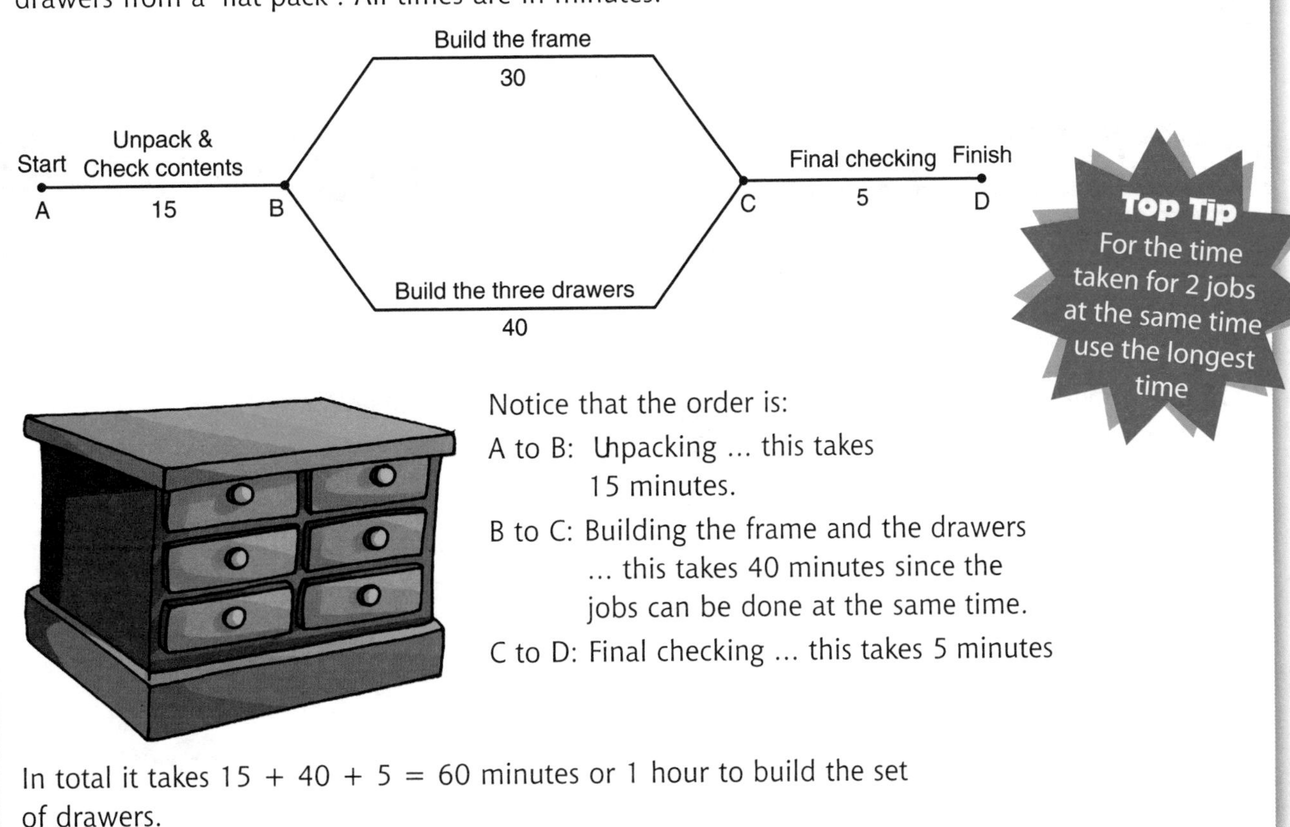

Notice that the order is:

A to B: Unpacking ... this takes 15 minutes.

B to C: Building the frame and the drawers ... this takes 40 minutes since the jobs can be done at the same time.

C to D: Final checking ... this takes 5 minutes

Top Tip
For the time taken for 2 jobs at the same time use the longest time

In total it takes 15 + 40 + 5 = 60 minutes or 1 hour to build the set of drawers.

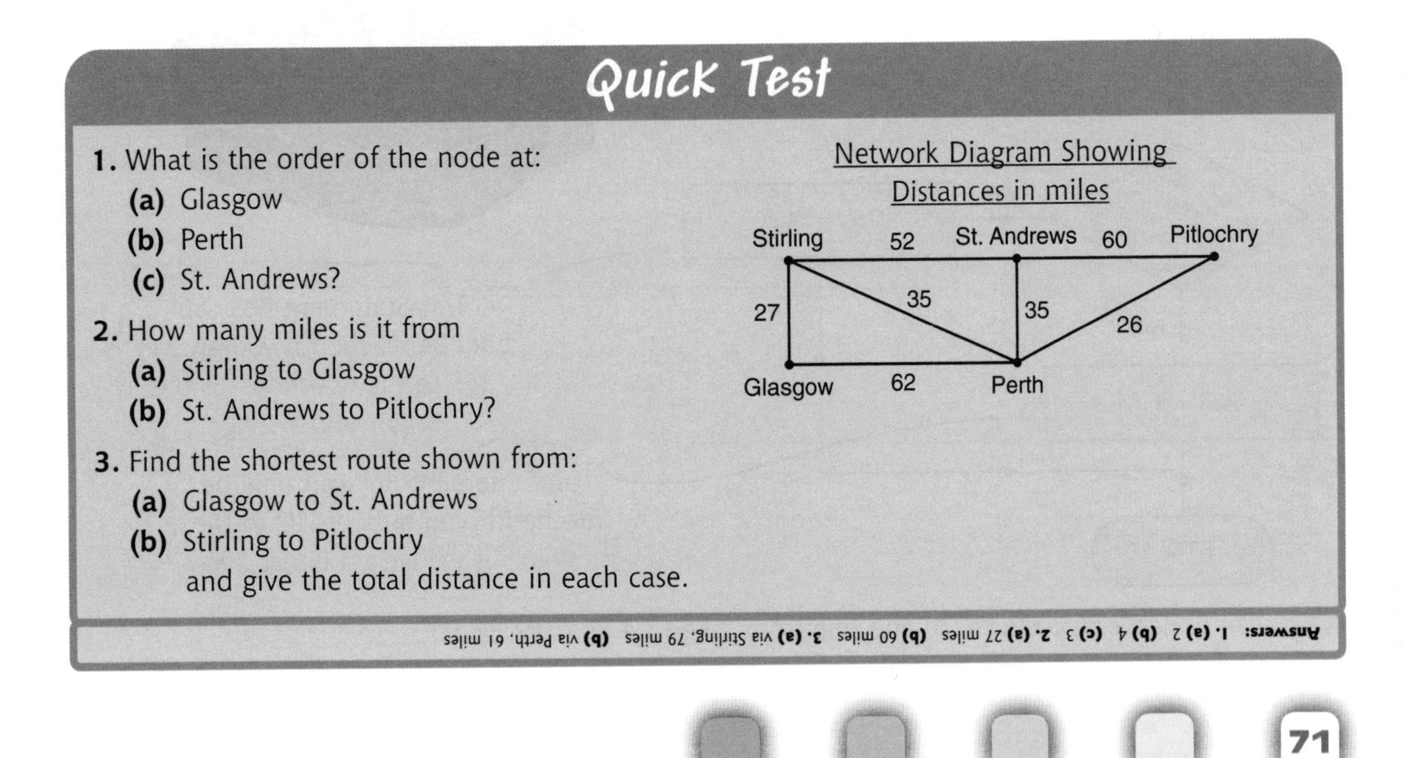

Quick Test

1. What is the order of the node at:
   (a) Glasgow
   (b) Perth
   (c) St. Andrews?

2. How many miles is it from
   (a) Stirling to Glasgow
   (b) St. Andrews to Pitlochry?

3. Find the shortest route shown from:
   (a) Glasgow to St. Andrews
   (b) Stirling to Pitlochry
   and give the total distance in each case.

Answers: 1. (a) 2 **(b)** 4 **(c)** 3 **2. (a)** 27 miles **(b)** 60 miles **3. (a)** via Stirling, 79 miles **(b)** via Perth, 61 miles

Flowcharts and Decision Tree Diagrams

How to read a Flowchart

Look first for the Start Box.... ...then follow the arrow

You might meet a Statement Box... ...this will give you a fact to use

or you might meet a Decision Box... Answer the question in the box and follow the YES or NO arrow depending on your answer.

When the arrows take you to the Stop Box.... ... you are finished

Top Tip
As you work through a Flowchart always write down the information you are given

An Example Flowchart

This flowchart allows you to calculate the cost of sending a parcel by post if it weighs no more than 4 kg.

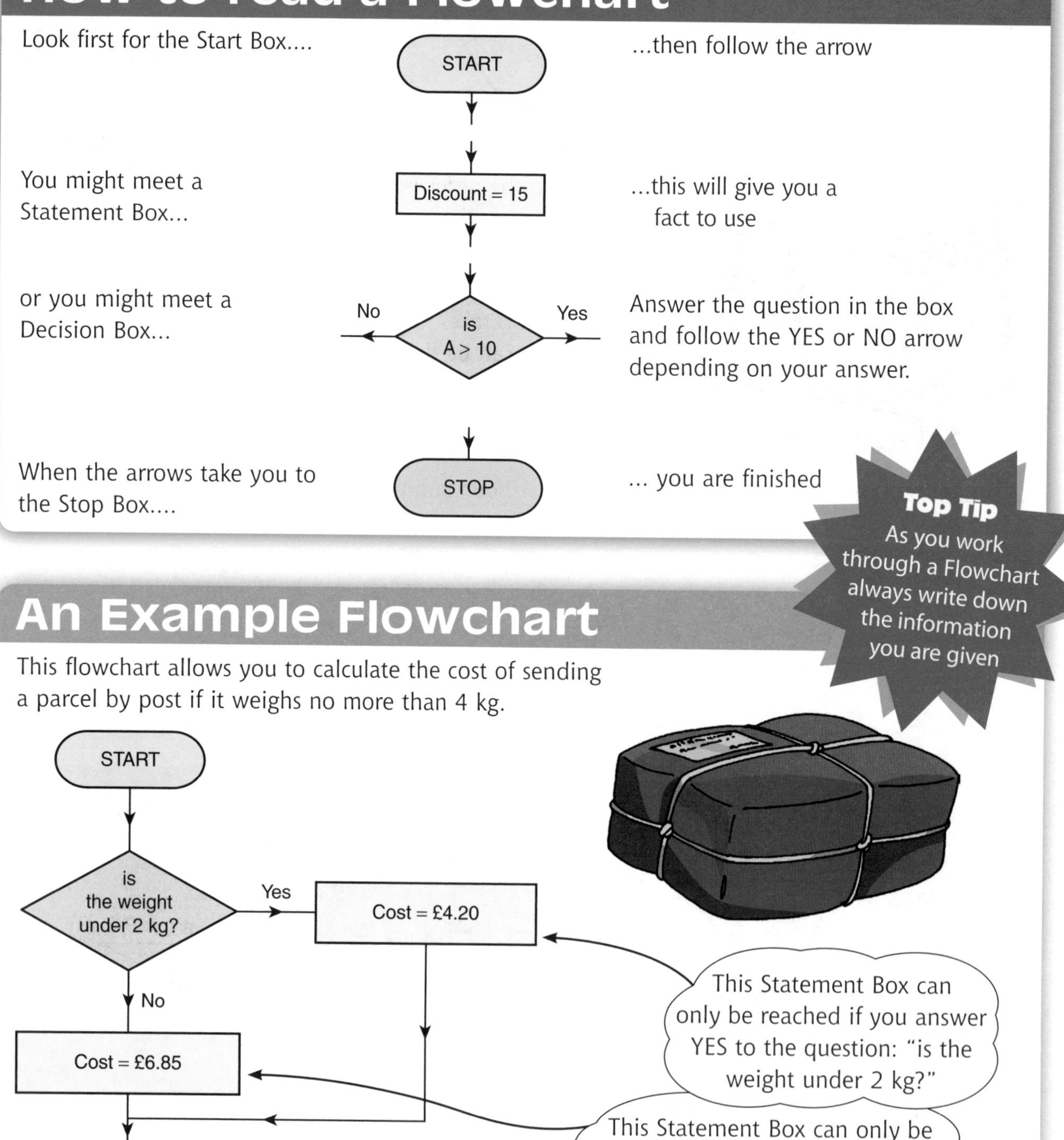

A parcel under 2 kg costs £4.20 and one from 2 kg to 4 kg costs £6.85.

Flowcharts with two Decision Boxes

In this example the flowchart helps a Garden Centre calculate the discount when they buy a large order of compost from their supplier.

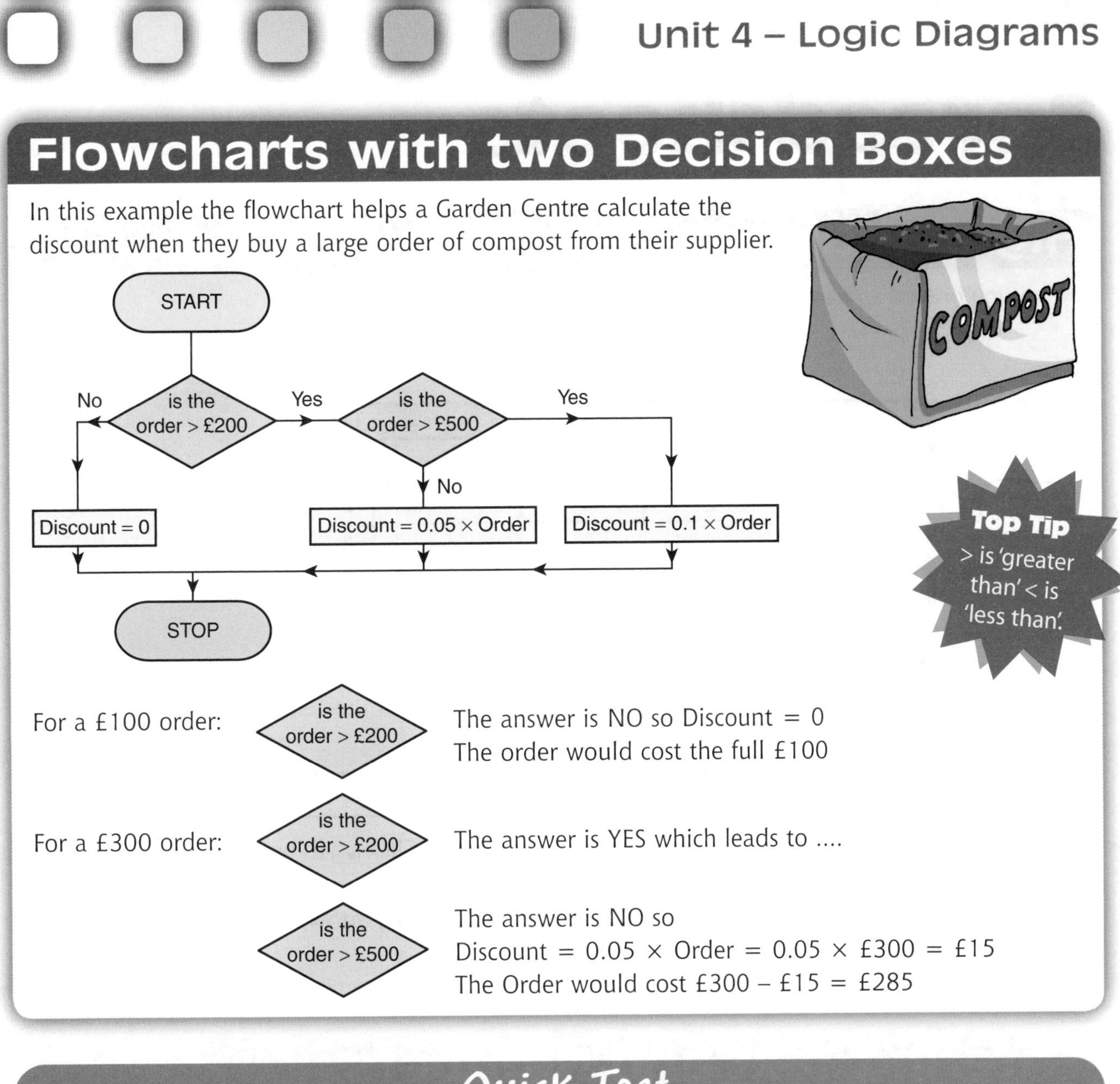

Top Tip
> is 'greater than' < is 'less than'.

For a £100 order: is the order > £200 — The answer is NO so Discount = 0
The order would cost the full £100

For a £300 order: is the order > £200 — The answer is YES which leads to ....

is the order > £500 — The answer is NO so
Discount = 0.05 × Order = 0.05 × £300 = £15
The Order would cost £300 – £15 = £285

Quick Test

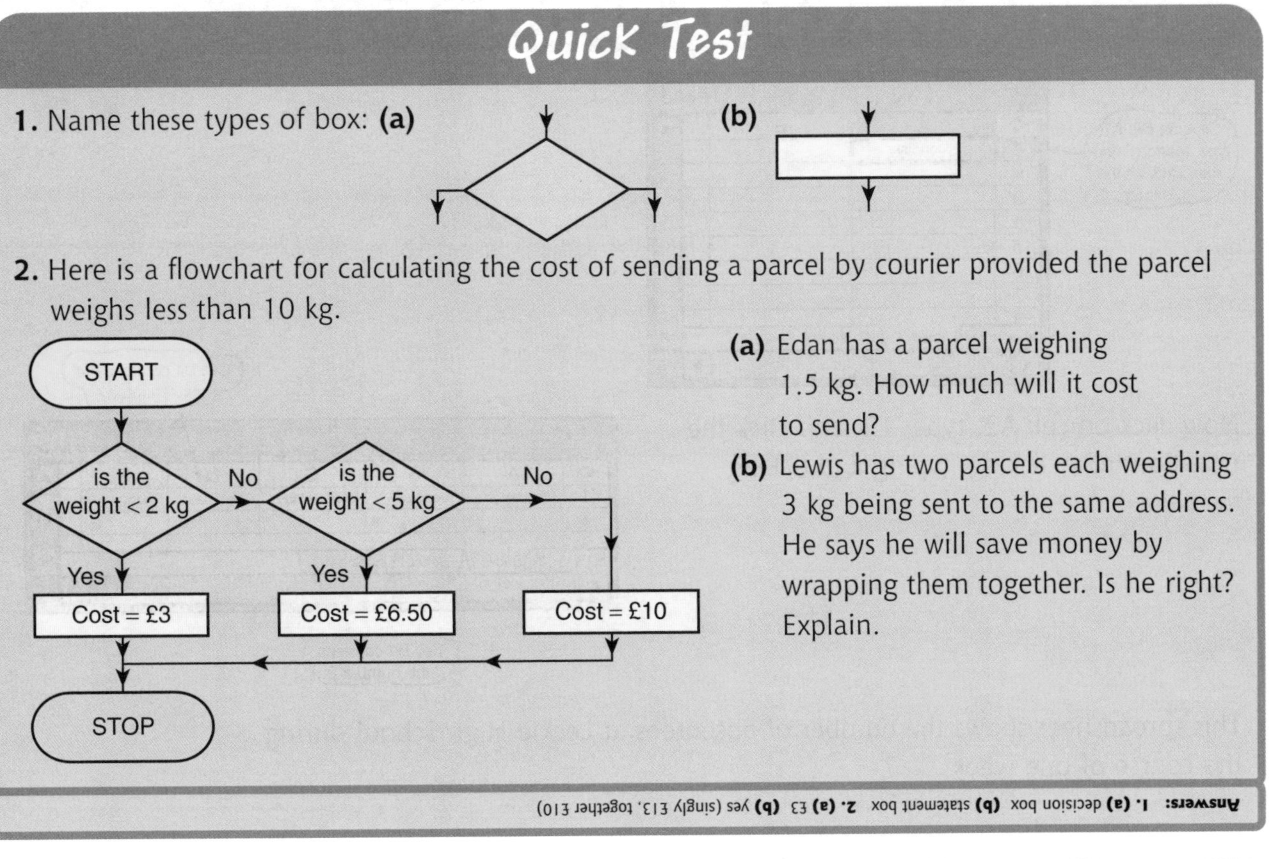

1. Name these types of box: **(a)** **(b)**

2. Here is a flowchart for calculating the cost of sending a parcel by courier provided the parcel weighs less than 10 kg.

(a) Edan has a parcel weighing 1.5 kg. How much will it cost to send?

(b) Lewis has two parcels each weighing 3 kg being sent to the same address. He says he will save money by wrapping them together. Is he right? Explain.

Answers: **1. (a)** decision box **(b)** statement box **2. (a)** £3 **(b)** yes (singly £13, together £10)

Spreadsheets

Highlighting Cells and the Edit Bar

When you open a spreadsheet page it looks like this:

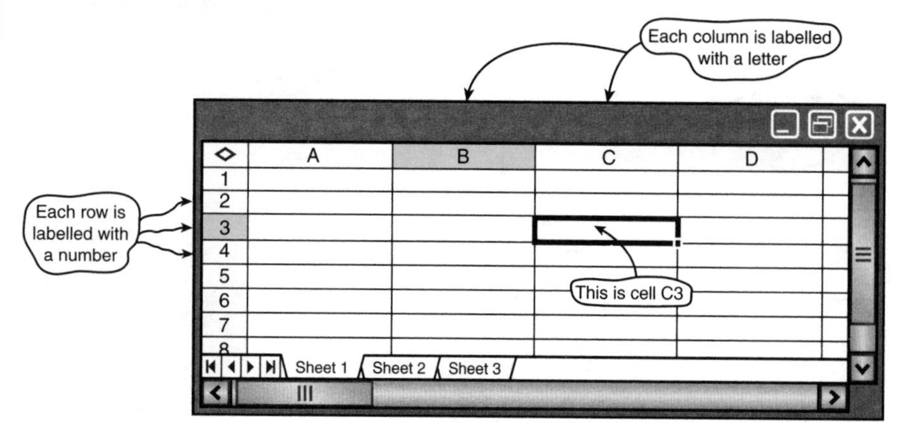

Clicking on a cell highlights that cell (like C3 above). The name of the cell, C3, appears above the spreadsheet in the EDIT BAR:

C3 ✕ ✓ =

Any typing you do in cell C3 will now appear in this Edit Bar. Click ☑ to enter your typing in the highlighted cell.

Different cells can be highlighted by clicking on them or using the arrow keys to move from cell to cell.

Top Tip
Letter for column
Number for row.

Entering Text and Numbers

Click on cell A1, type 'Monday' and then click on the tick ☑

	A	B
1	Monday	
2		
3		
4		
5		
6		
7		

Now click on cell A2, type '12' and click the tick. Continue clicking, typing and ticking to make:

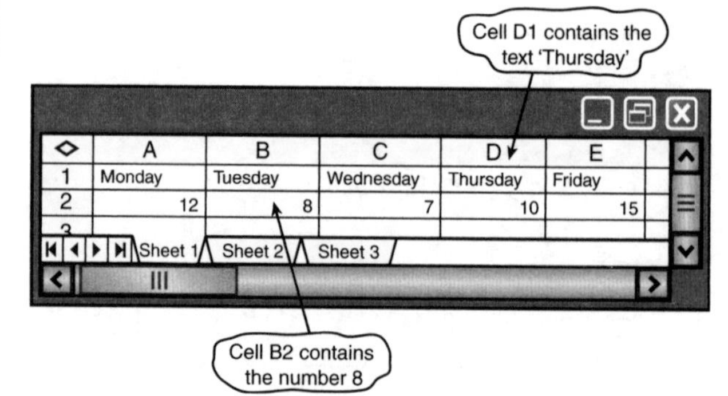

	A	B	C	D	E
1	Monday	Tuesday	Wednesday	Thursday	Friday
2	12	8	7	10	15

This spreadsheet shows the number of absentees at Leckie High School during the course of one week.

Entering Formulae

Click cell F2 and type:
"= A2 + B2 +
C2 + D2 + E2"
then click the tick ✓

F2 | = A2 + B2 + C2 + D2 + E2

	A	B	C	D	E	F
1	Monday	Tuesday	Wednesday	Thursday	Friday	Weekly Total
2	12	8	7	10	15	52
3						
4						
5						

Sheet 1 | Sheet 2 | Sheet 3

The result, 52, appears. Typing "=" means "what follows is a formula". The formula "= A2 + B2 + C2 + D2 + E2" adds up the numbers in these cells, in this case 12 + 8 + 7 + 10 + 15

Top Tip
Formulae may look slightly different on different computers.

Special Formulae

= SUM (A2:E2) — This formula adds together the numbers in cells A2, B2, C2, D2 and E2. For the spreadsheet above this gives 52.

= AVERAGE (A2:E2) — This calculates the average (mean) of the numbers in cells A2, B2, C2, D2 and E2. For the spreadsheet above this gives 10.4

= B2 * C2/D2 — *means multiply / means divide. For the spreadsheet above this gives 8 × 7 ÷ 10 = 5.6

Copying and Pasting Formulae

'Copy' cell C2 and 'paste' in cell C3. The formula changes from "= A2 * B2" to "= A3 * B3". The result will be £40.00 in cell C3.

C2 | = A2 * B2

	A	B	C	D	E	F
1	No. of Items	Cost of Item	Total			
2	3	£12.00	£36.00			
3	5	£8.00				
4	2	£17.00				
5						

Sheet 1 | Sheet 2 | Sheet 3

Quick Test

1. In the spreadsheet above which cell is the text 'Total' in?
2. The formula in cell C2 is pasted in cell C4. What result will appear in cell C4?
3. The grand total of all the items is to appear in cell C5. Complete the formula that will give this result: = SUM(:   )

Answers: 1. C1 2. = A4 * B4 3. = SUM(C2 : C4)

Scales and Scale Drawings

Measuring and Drawing Lines Accurately

Working with scale drawings you need to be very accurate:

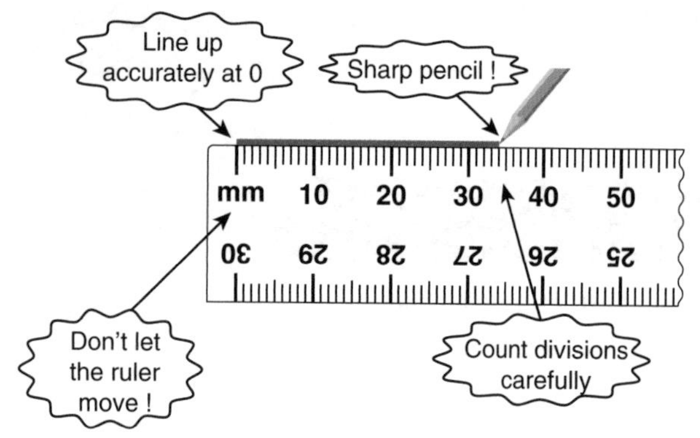

This diagram shows the things to take care over when drawing a line 3.4 cm long.

Apart from the 'Sharp Pencil' this advice holds for measuring lines also.

Top Tip
The more careful and accurate you are the more marks you'll gain.

Measuring and Drawing Angles Accurately

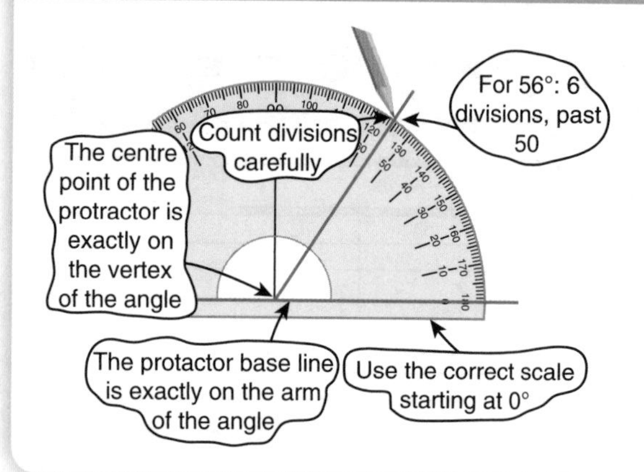

This diagram shows the things to take care over when measuring or drawing an angle of 56°.

When drawing the angle draw the base line first, place the protractor as shown then make an accurate mark at the tip of the pencil to draw the other arm.

Scale

Here is an example of a scale: 1 cm to 25 m or 1 cm : 25 m

Using this scale a 1 cm line on the map represents 25 m in real life.

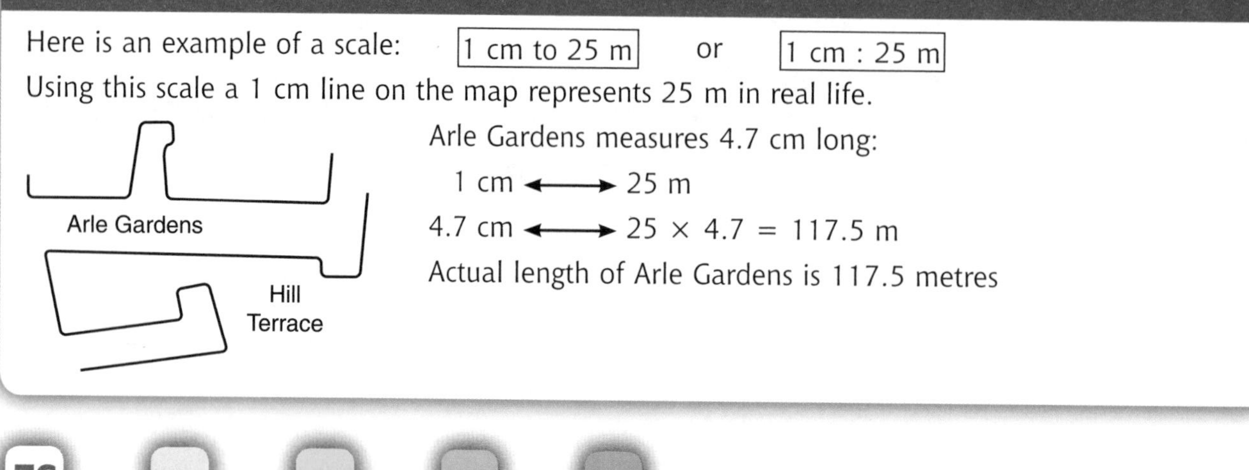

Arle Gardens measures 4.7 cm long:

1 cm ⟷ 25 m

4.7 cm ⟷ $25 \times 4.7 = 117.5$ m

Actual length of Arle Gardens is 117.5 metres

Angles of Elevation and Depression

This diagram shows the angle of elevation of the plane from the man.

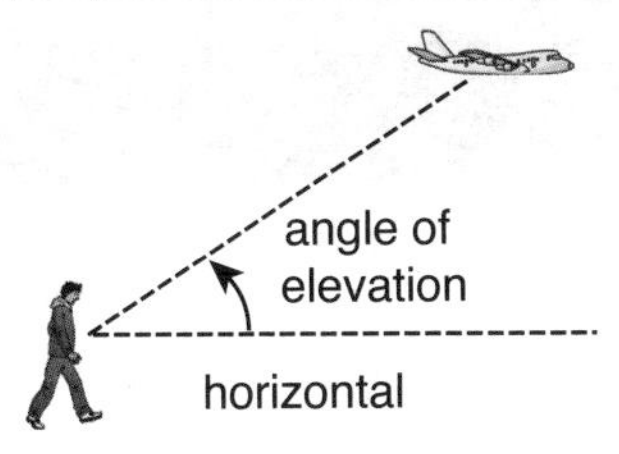

This diagram shows the angle of depression of the man from the plane.

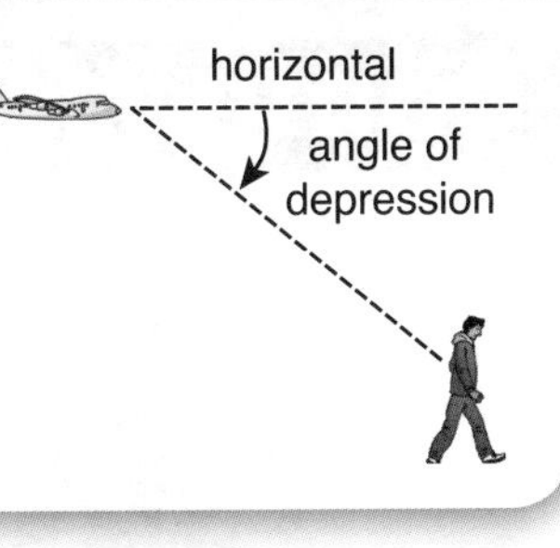

Constructing a Scale Drawing

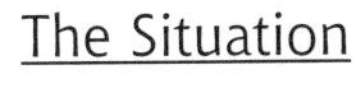

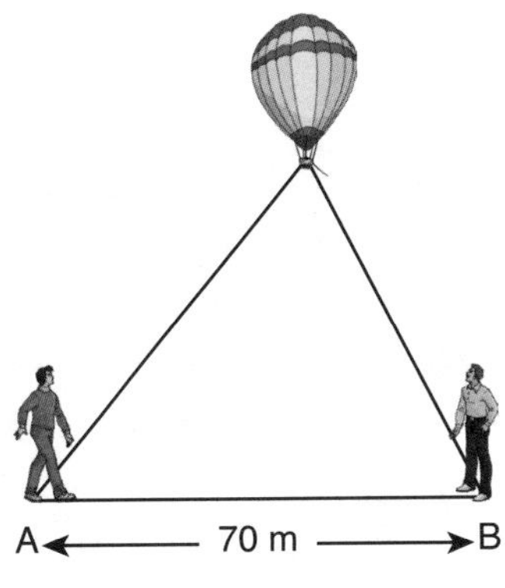

Two people are standing 70 metres apart looking up at a hot air balloon. The angle of elevation of the balloon is 50° from person A and 62° from person B as shown. Find the height of the balloon.

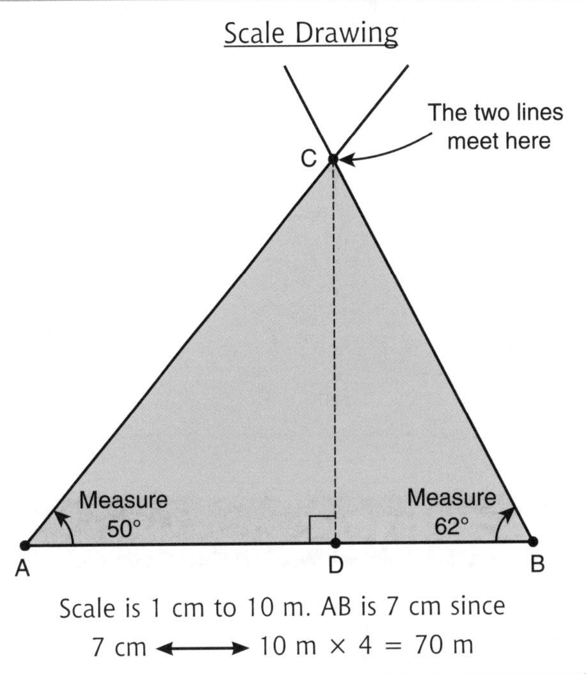

Scale is 1 cm to 10 m. AB is 7 cm since
7 cm ⟷ 10 m × 4 = 70 m

On the scale drawing line CD is at right-angles to AB and measures 5.1 cm.

In real life this is 5.1 × 20 m = 51 m.
This is the height of the balloon.

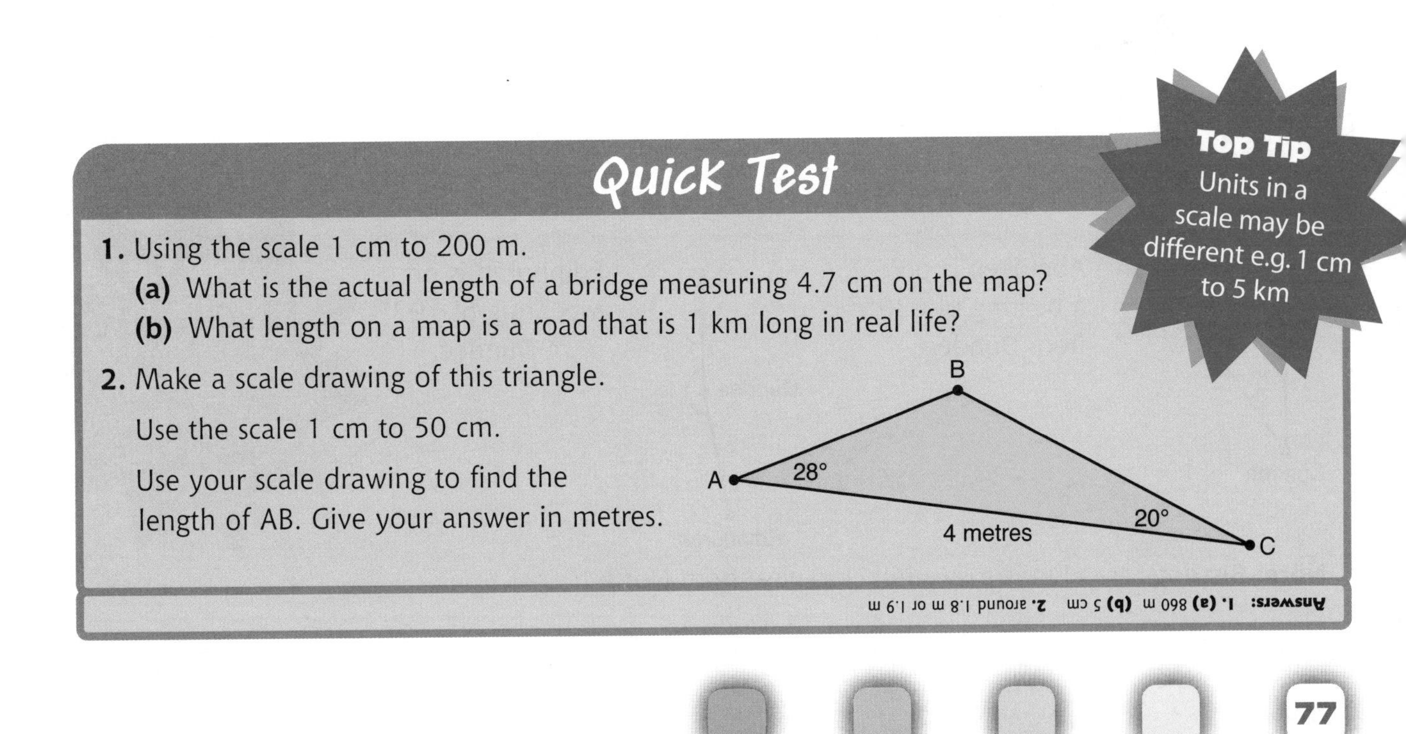

Quick Test

Top Tip
Units in a scale may be different e.g. 1 cm to 5 km

1. Using the scale 1 cm to 200 m.

(a) What is the actual length of a bridge measuring 4.7 cm on the map?

(b) What length on a map is a road that is 1 km long in real life?

2. Make a scale drawing of this triangle.

Use the scale 1 cm to 50 cm.

Use your scale drawing to find the length of AB. Give your answer in metres.

Answers: **1. (a)** 860 m **(b)** 5 cm **2.** around 1.8 m or 1.9 m

Directions and Bearings

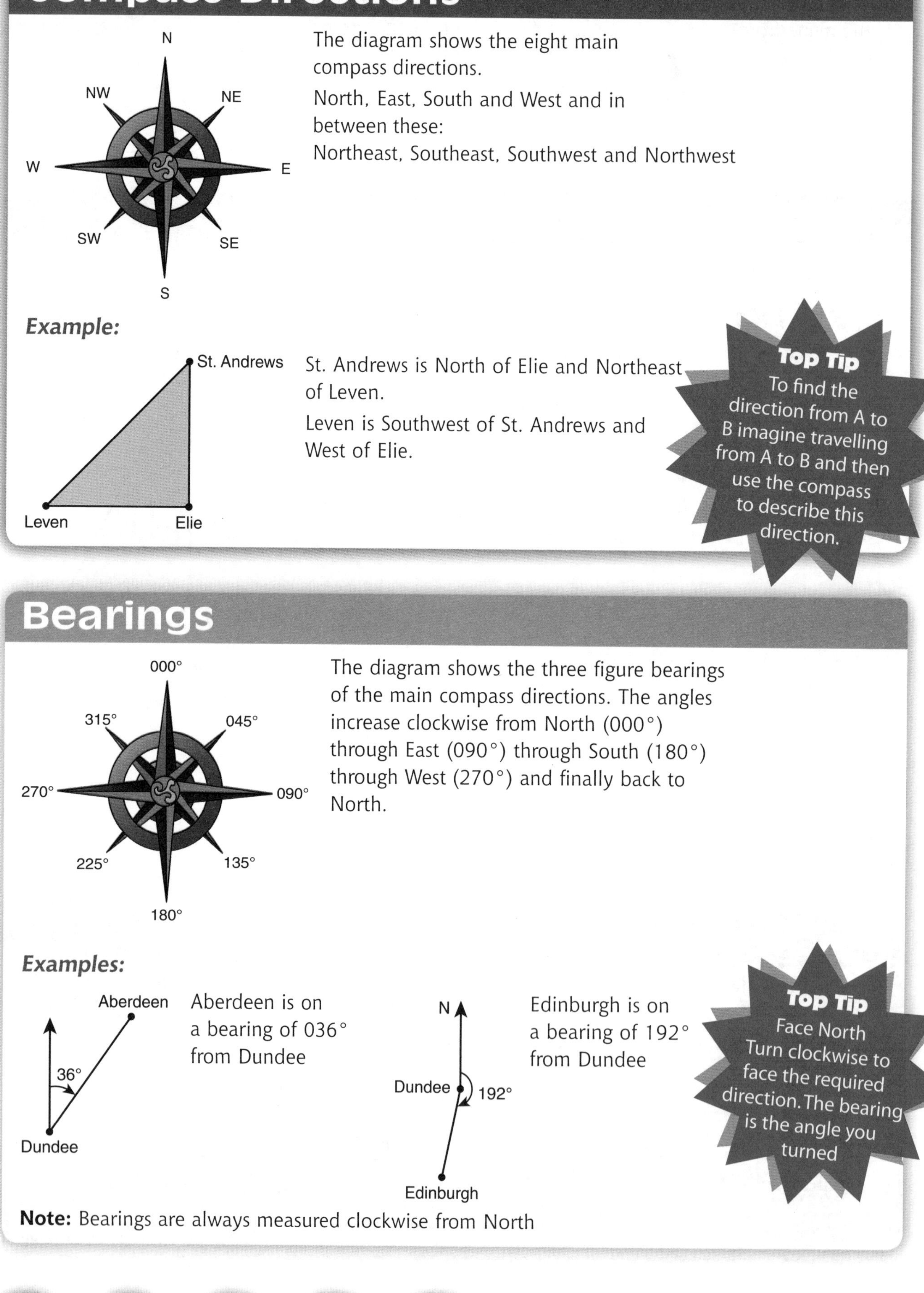

Compass Directions

The diagram shows the eight main compass directions.

North, East, South and West and in between these:
Northeast, Southeast, Southwest and Northwest

Example:

St. Andrews is North of Elie and Northeast of Leven.

Leven is Southwest of St. Andrews and West of Elie.

Top Tip
To find the direction from A to B imagine travelling from A to B and then use the compass to describe this direction.

Bearings

The diagram shows the three figure bearings of the main compass directions. The angles increase clockwise from North (000°) through East (090°) through South (180°) through West (270°) and finally back to North.

Examples:

Aberdeen is on a bearing of 036° from Dundee

Edinburgh is on a bearing of 192° from Dundee

Top Tip
Face North
Turn clockwise to face the required direction. The bearing is the angle you turned

Note: Bearings are always measured clockwise from North

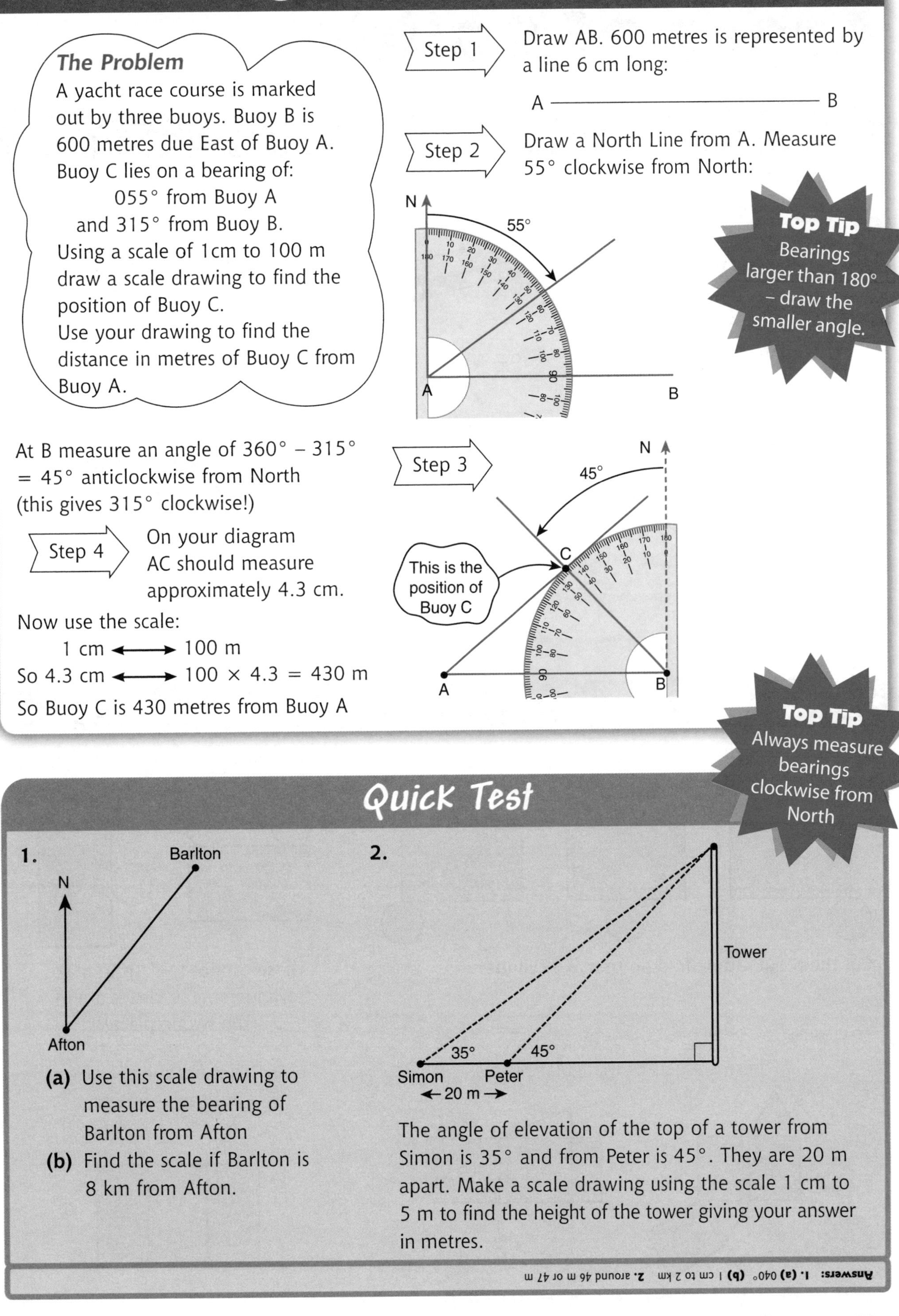

Constructing a Bearings Diagram

The Problem

A yacht race course is marked out by three buoys. Buoy B is 600 metres due East of Buoy A. Buoy C lies on a bearing of:

055° from Buoy A

and 315° from Buoy B.

Using a scale of 1cm to 100 m draw a scale drawing to find the position of Buoy C.

Use your drawing to find the distance in metres of Buoy C from Buoy A.

Step 1 Draw AB. 600 metres is represented by a line 6 cm long:

Step 2 Draw a North Line from A. Measure 55° clockwise from North:

Top Tip Bearings larger than 180° – draw the smaller angle.

Step 3 At B measure an angle of 360° – 315° = 45° anticlockwise from North (this gives 315° clockwise!)

Step 4 On your diagram AC should measure approximately 4.3 cm.

Now use the scale:

1 cm ⟷ 100 m

So 4.3 cm ⟷ 100 × 4.3 = 430 m

So Buoy C is 430 metres from Buoy A

Top Tip Always measure bearings clockwise from North

Quick Test

1.

(a) Use this scale drawing to measure the bearing of Barlton from Afton

(b) Find the scale if Barlton is 8 km from Afton.

2.

The angle of elevation of the top of a tower from Simon is 35° and from Peter is 45°. They are 20 m apart. Make a scale drawing using the scale 1 cm to 5 m to find the height of the tower giving your answer in metres.

Answers: 1. (a) 040° **(b)** 1 cm to 2 km **2.** around 46 m or 47 m

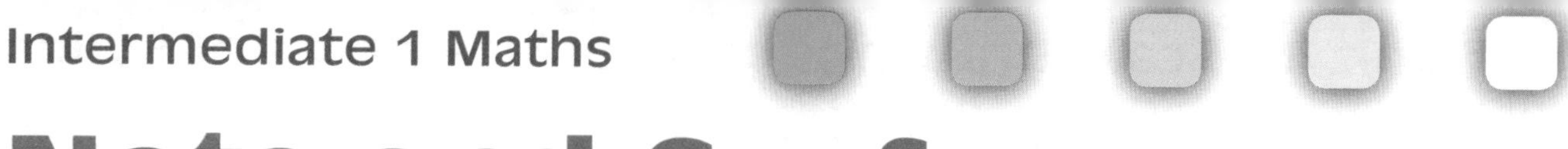

Nets and Surface Areas of Solids

Nets of Solid Shapes

The solid shape: The net:

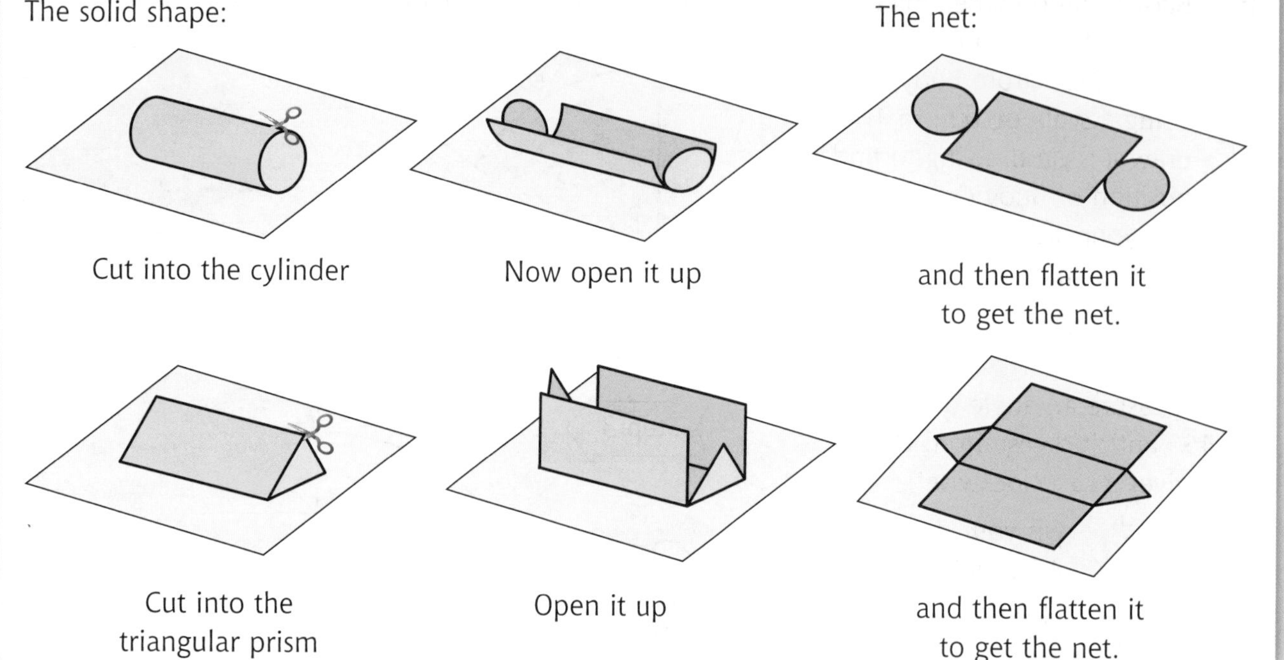

Cut into the cylinder

Now open it up

and then flatten it to get the net.

Cut into the triangular prism

Open it up

and then flatten it to get the net.

Is it a net or not?

These are all nets of cylinders:

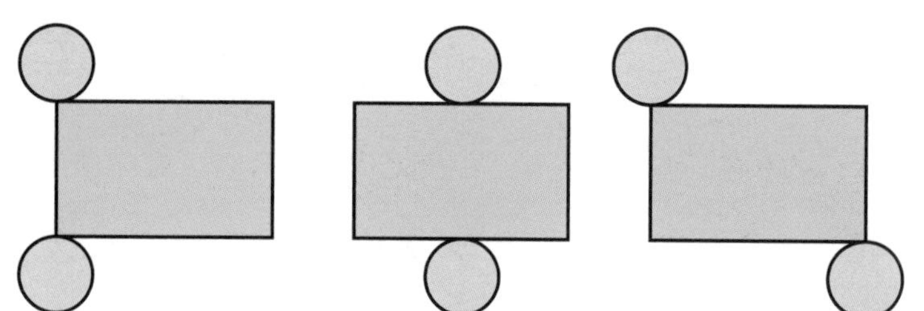

Cut them out – they fold up to make cylinders

These are not nets of cylinders:

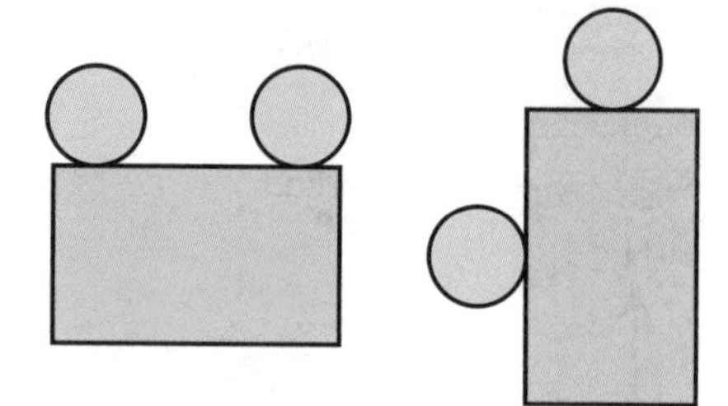

These cannot fold up to make cylinders – the circles are in the wrong place!

These are nets of triangular prisms:

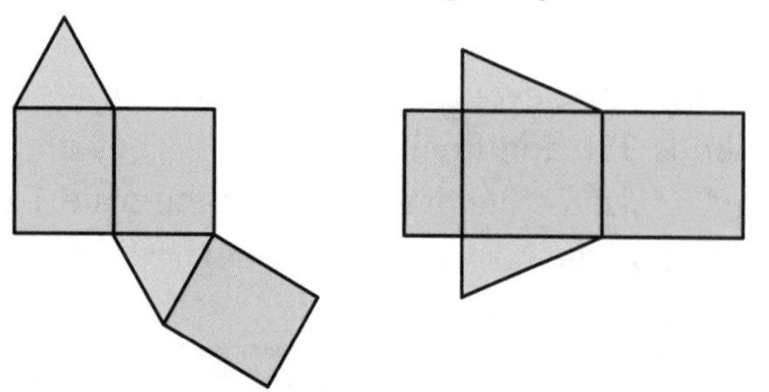

This is not the net of a triangular prism:

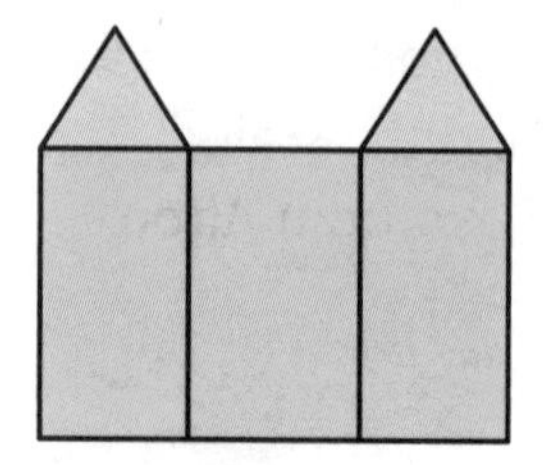

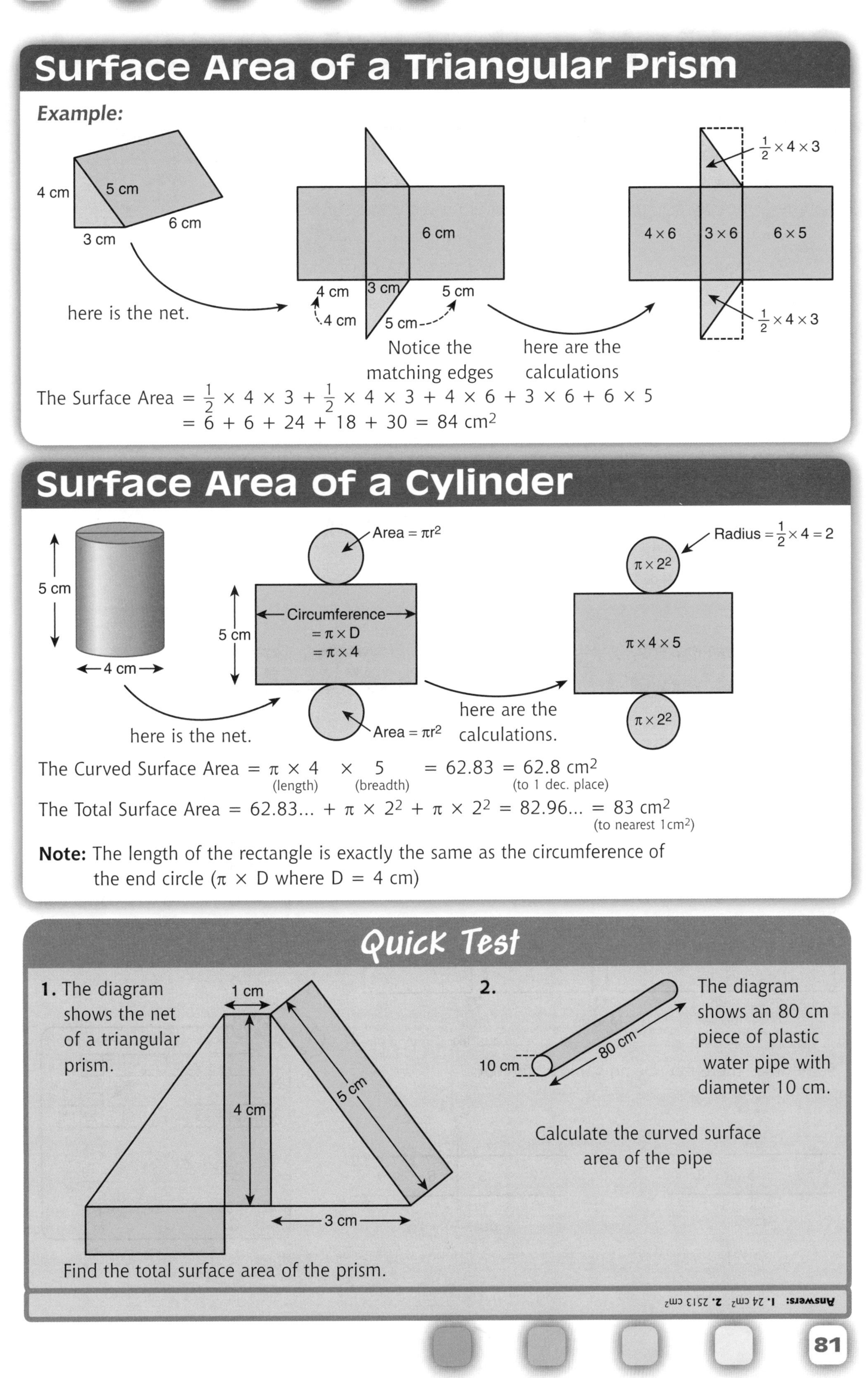

Surface Area of a Triangular Prism

Example:

here is the net.

Notice the matching edges

here are the calculations

The Surface Area $= \frac{1}{2} \times 4 \times 3 + \frac{1}{2} \times 4 \times 3 + 4 \times 6 + 3 \times 6 + 6 \times 5$

$= 6 + 6 + 24 + 18 + 30 = 84\ \text{cm}^2$

Surface Area of a Cylinder

here is the net.

here are the calculations.

The Curved Surface Area $= \pi \times 4$ (length) $\times\ 5$ (breadth) $= 62.83 = 62.8\ \text{cm}^2$ (to 1 dec. place)

The Total Surface Area $= 62.83... + \pi \times 2^2 + \pi \times 2^2 = 82.96... = 83\ \text{cm}^2$ (to nearest 1cm^2)

Note: The length of the rectangle is exactly the same as the circumference of the end circle ($\pi \times D$ where D = 4 cm)

Quick Test

1. The diagram shows the net of a triangular prism.

Find the total surface area of the prism.

2. The diagram shows an 80 cm piece of plastic water pipe with diameter 10 cm.

Calculate the curved surface area of the pipe

Answers: **1.** 24 cm² **2.** 2513 cm²

Statistical Assignment

Outline Structure

You are required to write a short statistical assignment. Your teacher will give you help and advice on a topic that you could investigate. Choose a topic that you find interesting. Here is an outline structure for your final report:

- An introduction where you explain the aims of your project and how you gathered your data.
- List the sets of data that you collected. There are usually two sets.
- Present the data in a more meaningful way for example: a graph, a table or a stem-and-leaf diagram.
- Calculate statistics using your data: minimum and maximum values, lower and upper quartiles and the median (i.e. a five-figure summary); also the range and interquartile range.
- Construct a boxplot for each set of data.
- Compare your data sets by discussing the boxplots. Always back up your statements with the statistics you calculated.
- A conclusion where you summarise your findings.

Statistics – Calculating Quartiles

If a data set is arranged in order (smallest to largest) and written as a list on a piece of tape, the tape can be cut into four equal pieces:

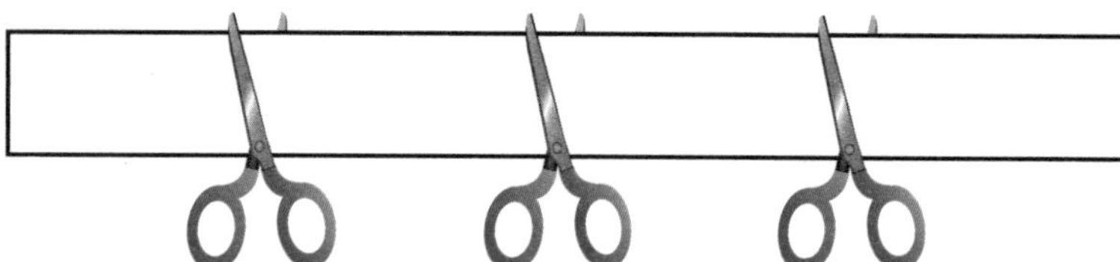

The values in the data set at the places where the tape is cut have names:

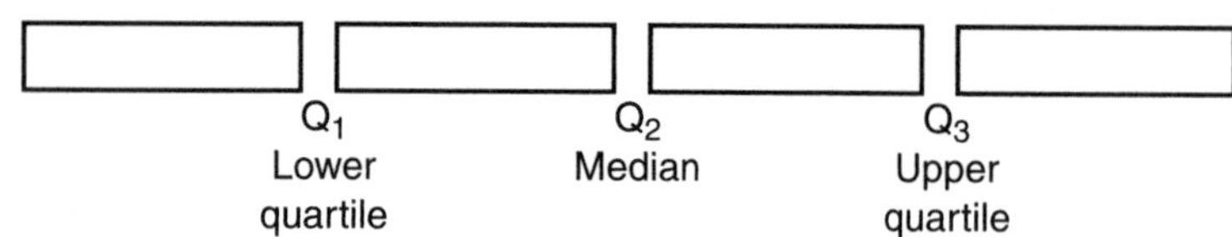

Sometimes there may be no value at the cut:

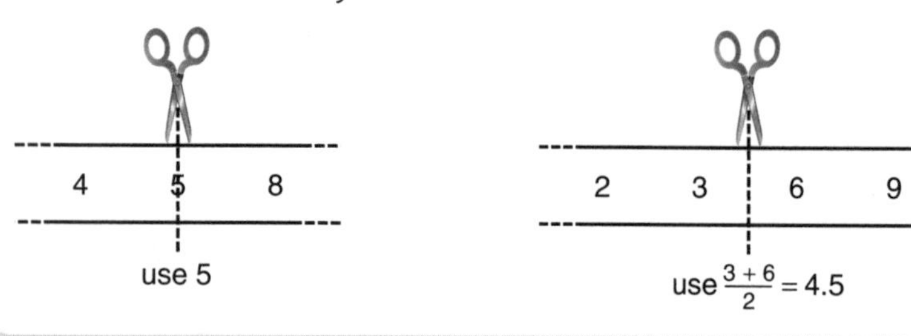

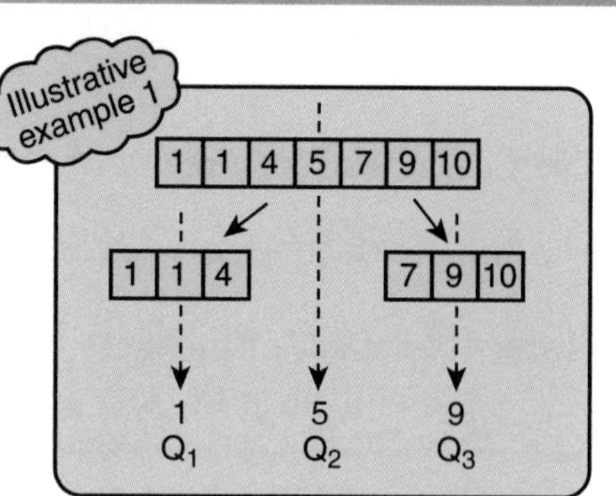

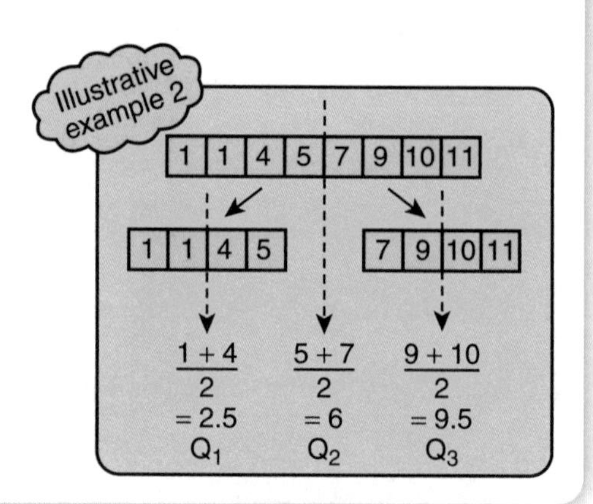

Statistics – Calculating the Range & Interquartile Range

The Range tells you how far apart the smallest and largest values are:

$$\text{Range} = \text{Greatest Value} - \text{Least Value}$$

The Interquartile Range tells you how far apart the Upper and Lower Quartiles are:

$$\text{Interquartile Range} = \text{Upper Quartile}(Q_3) - \text{Lower Quartile }(Q_1)$$

Boxplots

Boxplots are an effective way to illustrate the greatest and least values from a data set, along with its Median and Upper and Lower Quartiles:

Note: The 'Box' stretches from Q_1 to Q_3 so the length of the Box is the same as the Interquartile Range (in this case 3)

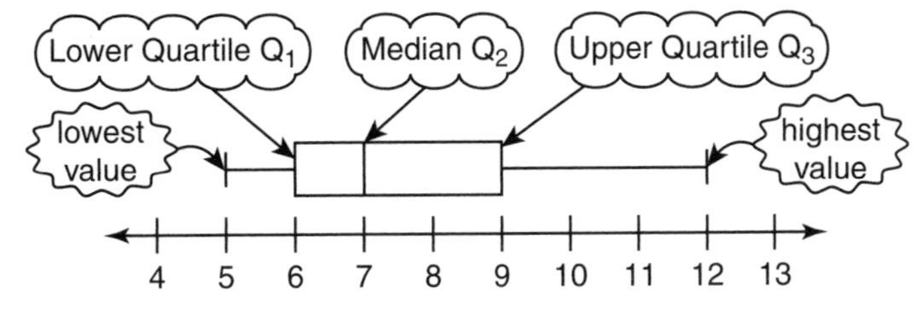

Note: The two 'whiskers' sticking out each end of the 'Box' stretch from the least value (5) to the greatest value (12). The length between the two ends of the whiskers (in this case 7) is the same as the Range.

Comparing Datasets

Example:

Each football season 12 teams compete in the Scottish Premier League (Division 1). Here are two boxplots showing the total goals scored by each of the 12 teams during season 2006–07 and season 2007–08:

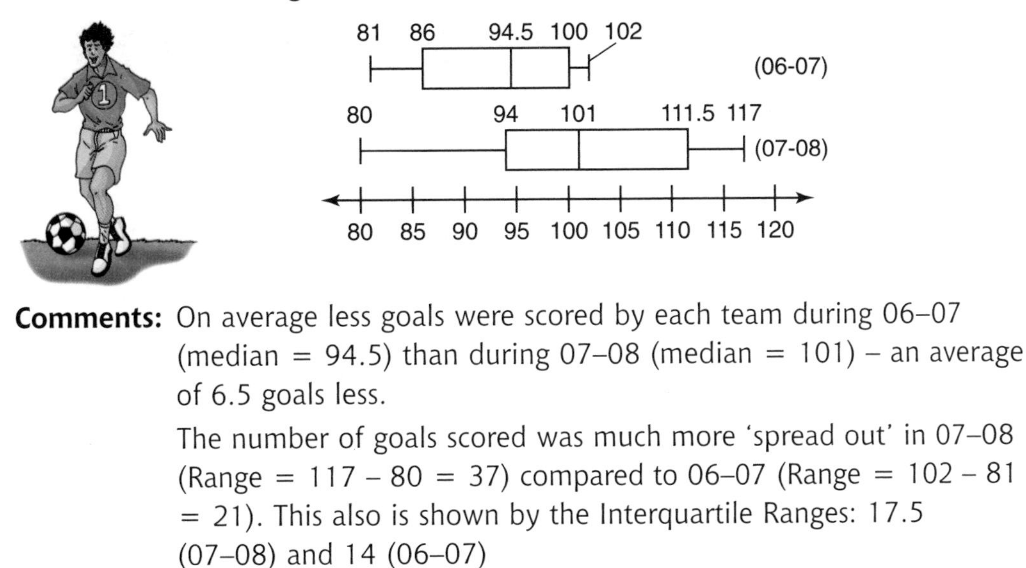

Comments: On average less goals were scored by each team during 06–07 (median = 94.5) than during 07–08 (median = 101) – an average of 6.5 goals less.

The number of goals scored was much more 'spread out' in 07–08 (Range = 117 – 80 = 37) compared to 06–07 (Range = 102 – 81 = 21). This also is shown by the Interquartile Ranges: 17.5 (07–08) and 14 (06–07)

Practice Unit 4 Test

Outcome 1

1. Abi's payslip is shown. Her basic rate of pay is £11 per hour. And she is paid time and a half for overtime.
   (a) For the week shown she worked 6 hours overtime. Find her overtime pay and write it on her payslip
   (b) Now complete her payslip

Name	Employee No.	Week	Tax Code	NI Number
Abi Parker	328	12	K521	YT221233E
Basic Pay	Overtime	Bonus		Gross Pay
£440		—		
National Insurance	Income Tax	Pension		Total Deductions
£45.60	£83.32	—		
				Net Pay

2.

With Loan Protection

Loan Period	Amount of Loan	Monthly Repayment
6 months	£500	£93.20
	£1000	£186.41
12 Months	£500	£47.98
	£1000	£95.96
	£2000	£191.92

Without Loan Protection

Loan Period	Amount of Loan	Monthly Repayment
6 months	£500	£86.30
	£1000	£172.60
12 Months	£500	£44.43
	£1000	£88.85
	£2000	£177.70

Walter Howard took out a loan of £500 to be repaid over 12 months without loan protection.
   (a) Write down his monthly repayment.
   (b) Calculate the total amount he has repaid at the end of the 12 months.

Outcome 2

3. The diagram shows some routes and distances in miles between 5 places in Scotland.
   (a) Write down the order of the node at Oban.
   (b) How many miles is the shortest route from Oban to Pitlochry?

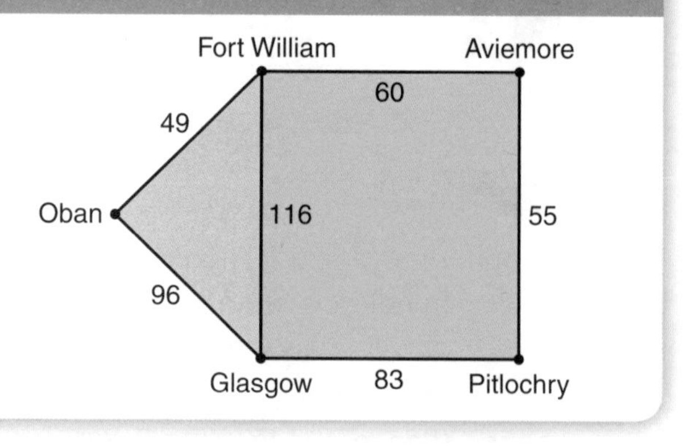

4. This flowchart shows how to calculate the cost of a taxi journey in Glasgow:

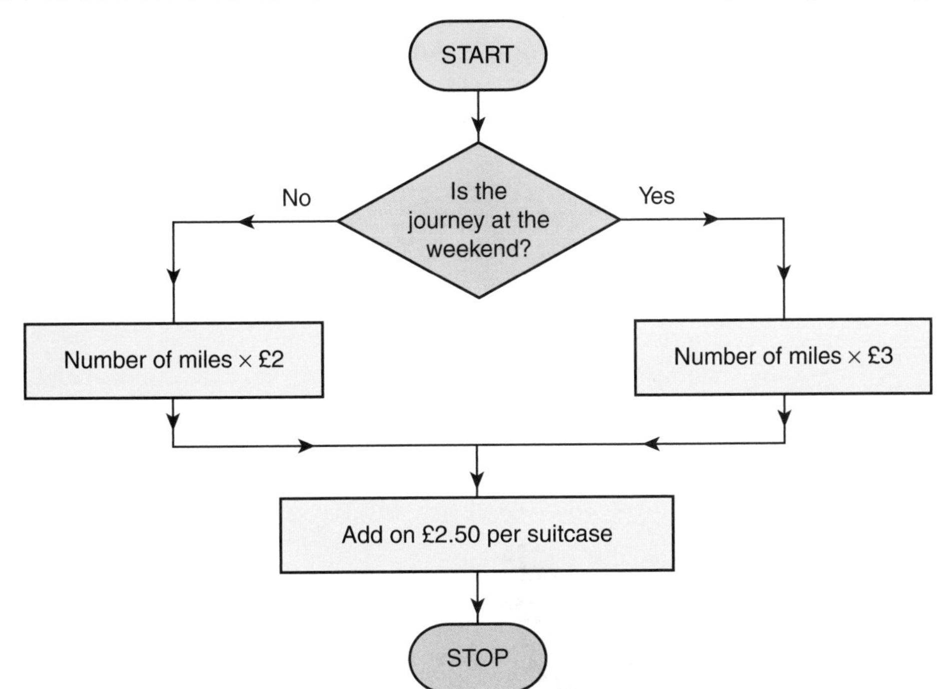

Mr Henderson and his two suitcases arrive by train in Glasgow on Saturday morning. How much will his 4 mile taxi journey cost?

5. The following is a record of the number of hours Donald spent on-line downloading data files at his work:

3 hours on Monday at £7.50 per hour
5 hours on Tuesday at £5.30 per hour
2 hours on Wednesday at £8 per hour
1 hour on Thursday at £4.80 per hour

(a) Enter this data into rows 2, 3, 4 and 5 of this spreadsheet grid:

	A	B	C	D	E
1		Day	No of Hours	Cost per hour	
2					
3					
4					
5					
6					

(b) In cell E1 write the heading: Daily Total

(c) Write down the amount which will appear in cell E2 when the formula = C2 * D2 is typed into cell E2.

(d) The formula in cell E2 is copied down into cells E3, E4 and E5. Complete this formula which puts the total cost of all the hours into cell E6

= SUM (:   )

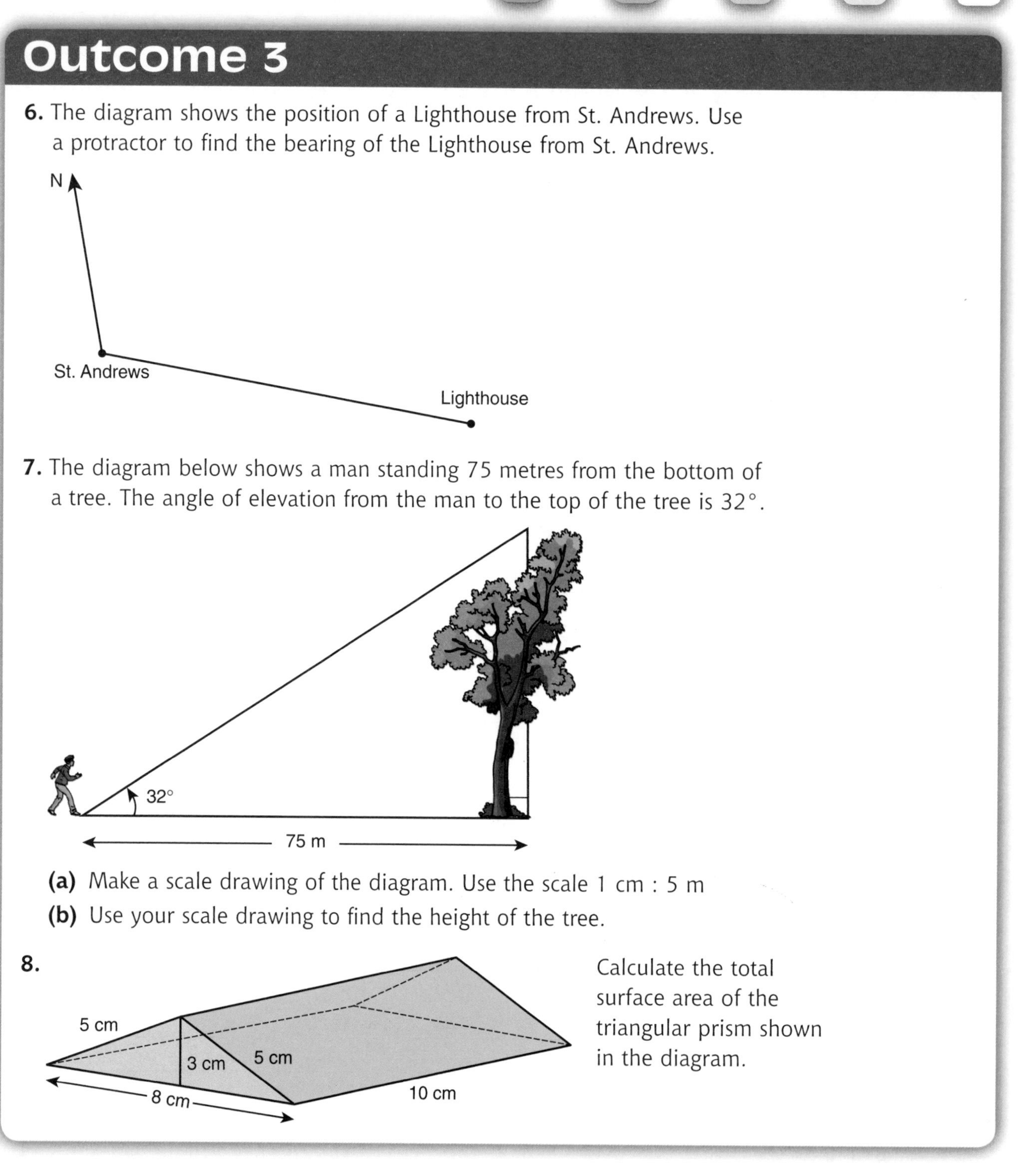

Outcome 3

6. The diagram shows the position of a Lighthouse from St. Andrews. Use a protractor to find the bearing of the Lighthouse from St. Andrews.

7. The diagram below shows a man standing 75 metres from the bottom of a tree. The angle of elevation from the man to the top of the tree is 32°.

(a) Make a scale drawing of the diagram. Use the scale 1 cm : 5 m

(b) Use your scale drawing to find the height of the tree.

8. Calculate the total surface area of the triangular prism shown in the diagram.

9. Two of the following diagrams represent the net of a cylinder (drawn to scale). Which two?

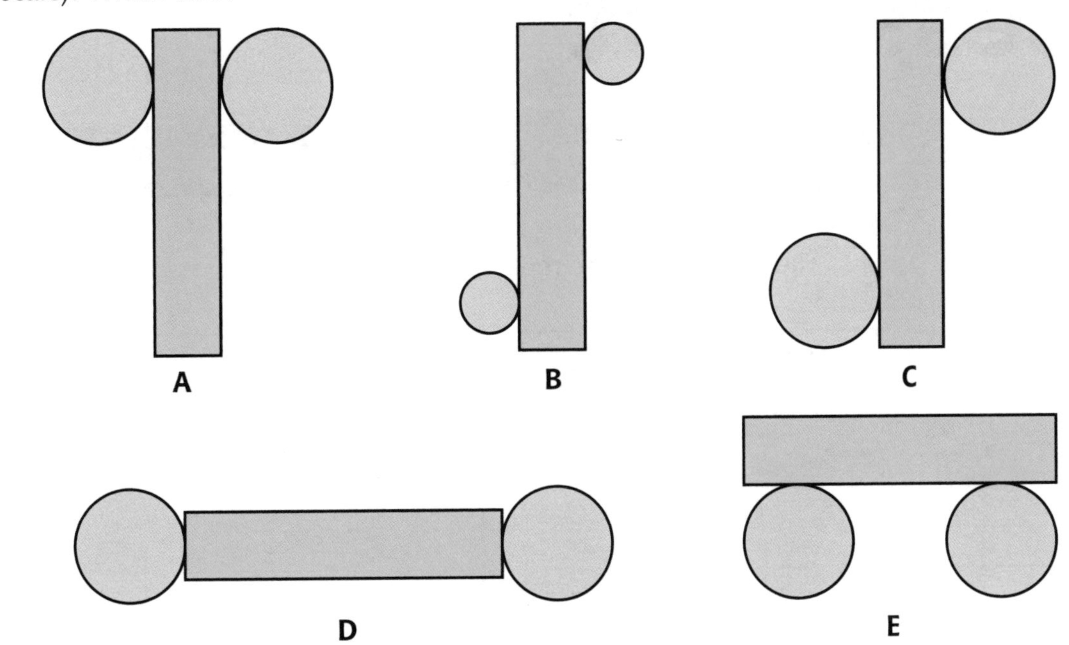

10. Two of the following diagrams represent the net of a triangular prism (drawn to scale). Which two?

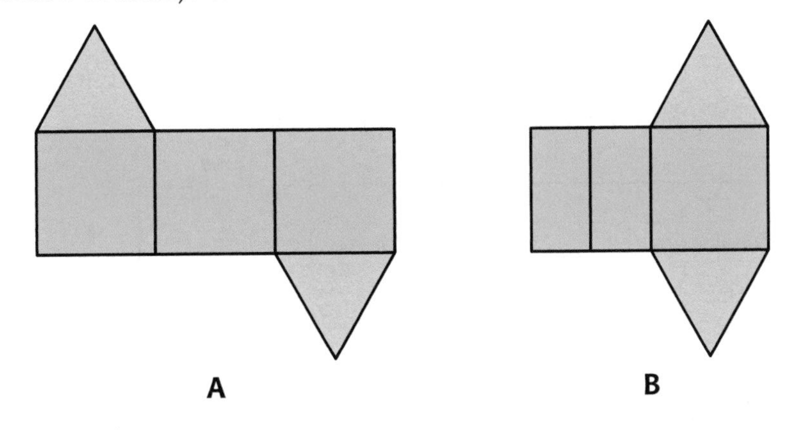

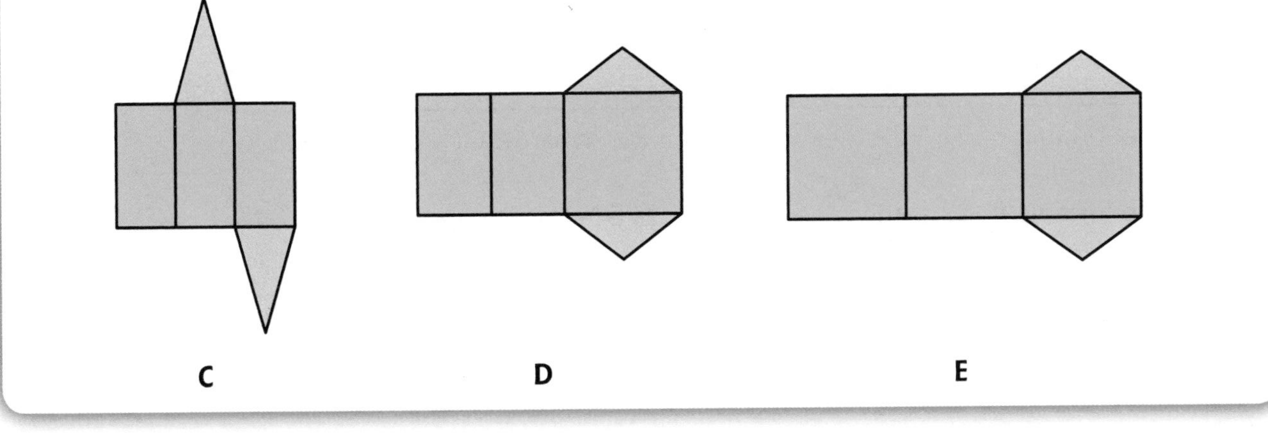

Preparation for Assessment

Practice Unit 4 Exam Questions

1. The table below shows the monthly payments to be made when money is borrowed from a bank. The table shows the repayments with and also without payment protection.

	1 Year		2 Years		3 Years	
Amount borrowed	Without payment protection	With payment protection	Without payment protection	With payment protection	Without payment protection	With payment protection
£500	£45.42	£46	£24.30	£24.98	£16.54	£17.25
£1000	£90.83	£92.08	£48.60	£49.96	£33.08	£34.51
£1500	£136.25	£138.13	£72.90	£74.94	£49.63	£51.76
£2000	£181.67	£184.17	£97.20	£99.92	£66.17	£69.02

(a) Mr Jacobs borrows £1500 over 2 years without payment protection. State his monthly payment.

(b) If he had taken out payment protection how much extra would he have paid in total over the 2 years?

2. A school uses a spreadsheet to calculate daily income from the school canteen.

	A	B	C	D
1	Item	Price of Item (£)	Number of Items	Total Cost
2	Jacket Potato	0.75	10	
3	Jacket Filling	0.50	15	
4	Filled Roll	1.15	20	
5	Pudding	0.55	18	
6	Fresh Fruit	0.45	12	
7				
8				

(a) The formula = B2 * C2 is entered in cell D2. What result would appear in the cell?

(b) The formula in D2 is then copied into cells D3, D4, D5 and D6. What formula is now in cell D6?

3. This network diagram shows the distances, in miles, between 5 towns and cities in Scotland.

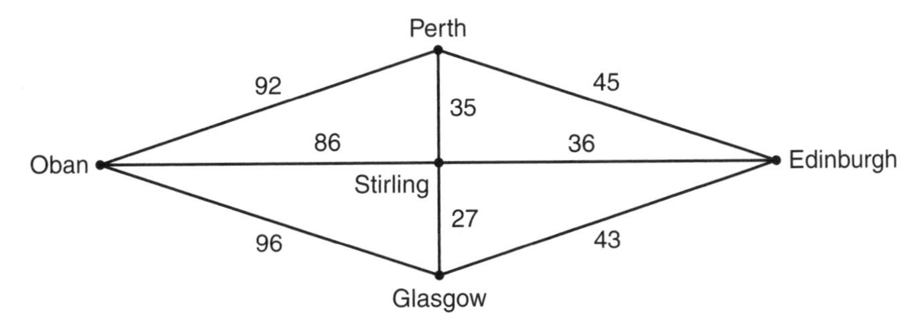

(a) State the order of the node at Stirling

(b) What is the length of the shortest route from Oban to Edinburgh?

4.

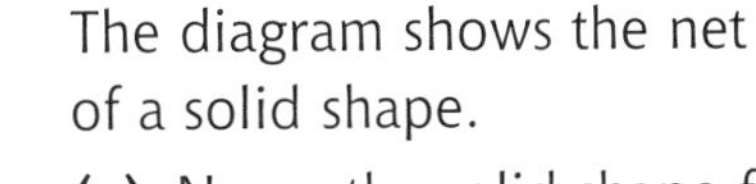

2 cm 4 cm 2 cm 3 cm 5 cm

The diagram shows the net of a solid shape.

(a) Name the solid shape formed from this net.

(b) Calculate the surface area of the solid shape.

5. These two box plots show the birth weights, in kg, of babies suffering a particular disease. Some survived and some died.

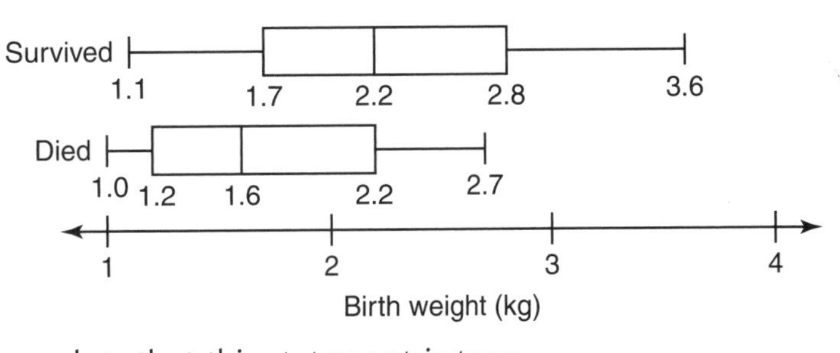

(a) Calculate the interquartile range for the babies that survived.

(b) Explain how you can tell from the box plots that this statement is true: "On average, the birth weight of the babies that survived was greater than those that died but their weights tended to be more variable".

6. Mrs Robbins sells computers. This flowchart is used to calculate her monthly salary in pounds.

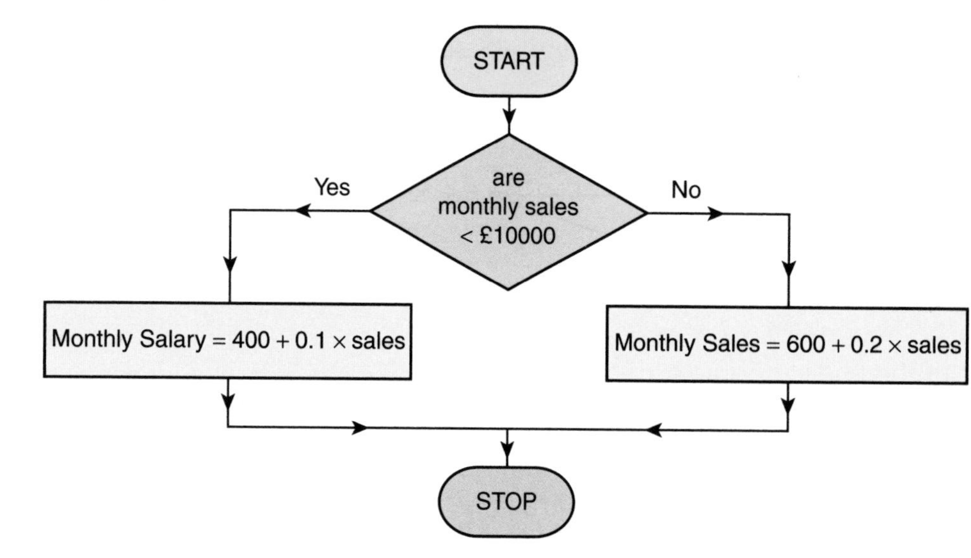

In January she sold £12000 worth of computers. Calculate her salary for January.

7.

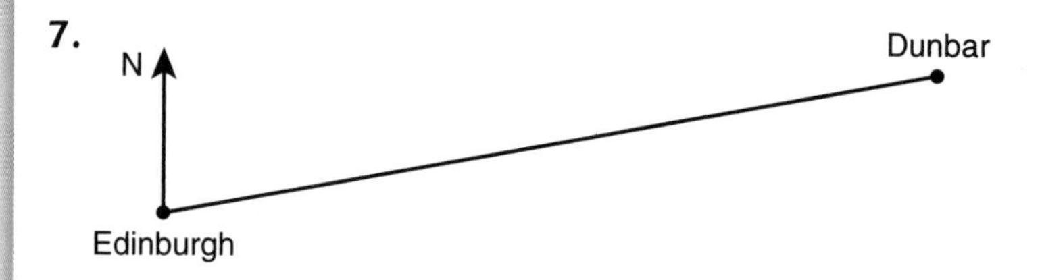

The above scale drawing shows the positions of two towns, Edinburgh and Dunbar, which are 40 km apart.

(a) Find the scale of the drawing

(b) A third town, Galashiels, is on a bearing of 150° from Edinburgh and on a bearing of 205° from Dunbar. Complete the scale drawing to show the position of Galashiels.

(c) Use your completed scale drawing to find the actual distance from Edinburgh to Galashiels. Give your answer in km.

8. Ian Hooper works as a mechanic. His basic rate of pay is £5.20 per hour for a 34 hour week. His overtime rate of pay is time and a half. Complete his payslip below for a week in which he works 7 hours overtime.

Payments				**Deductions**	
	Hours	Rate	Amount		Amount
Basic:	34	£5.20	£176.80	Tax:	£50.90
Overtime:	7			National Insurance:	£23.60
		Gross Pay:		Total Deductions:	
				Net Pay:	

9.

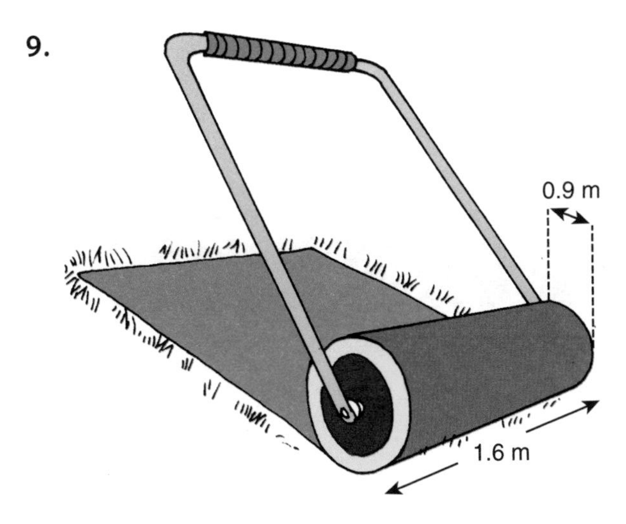

The diagram shows the area flattened by a grass roller after one complete turn of the drum. The drum has diameter 0.9 m and is 1.6 m wide.

Calculate the flattened area of grass. Give your answer correct to 1 decimal place.

10. The percentage scores obtained by pupils in class 4A2 in their term Exam in Mathematics were as follows:

60, 84, 68, 56, 56, 62, 84, 75, 63, 60, 55, 60, 59, 77

Complete this boxplot for this data by entering the three missing numbers

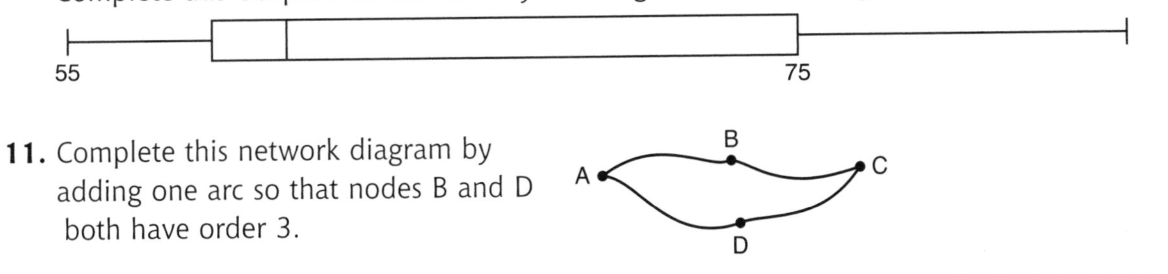

11. Complete this network diagram by adding one arc so that nodes B and D both have order 3.

Answers to Unit Tests

Unit 1 Answers

Outcome 1

1. £14.46 **2.** 38 p **3. (a)** 380 km **(b)** 32 circuits

Outcome 2

4. 5 m^2 **5. (a)** 16000 cm^3 (or 16 litres) **(b)** 27 cm^3

6. 28.3 m (to 1 dec pl) **7.** 28.3 cm^2 (to 1 dec pl)

Outcome 3

8. 20 **9.** £27.70 **10.** 120

Outcome 4

11. £218.04 **12.** £124 **13.** 8631 yuan

Unit 2 Answers

Outcome 1

1. (a) A(−3,0), B(−4, −3) **(b)** C:2 down on *y*-axis; D: 3 along 2 down

2. (a) 7 **(b)** −6 **(c)** 10

Outcome 2

3. (a) 30 miles **(b)** 20 minutes **(c)** steeper sloping line

4. 135 km **5.** $45\frac{1}{2}$ hours

Outcome 3

6. 7.8 m (to 1 dec pl)

Outcome 4

7. (a) 82%, 41% **(b)**

Stem	Leaf
4	1
5	8 9
6	0 2 3 4 5 8
7	0 1
8	2

(c) most pupils had a percentage score in the sixties

8. 20

9. (a)

Score	Tally	Frequency
0	卌 I	5
1	卌 IIII	9
2	II	2
3	I	1
4	II	2

(b) Not true (only 5 out of 20)

10. (a) (b)

(c) 11

Outcome 5

11. (a) 8.8 kg **(b)** 7 kg

(c) 7.5 kg **(d)** 9 kg

12. $\frac{1}{3}$

Unit 3 Answers

Outcome 1

1. 5 **2. (a)** 5m − 10 **(b)** 5y − 3 **3.** $6(x + 3)$

4. (a) $n = 19$ **(b)** $y = \frac{4}{3}$ or $1\frac{1}{3}$ **5. (a)** $k > 7$ **(b)** m < 6

Outcome 2

6. (a) −1, 1, 3, 5 **(b)**

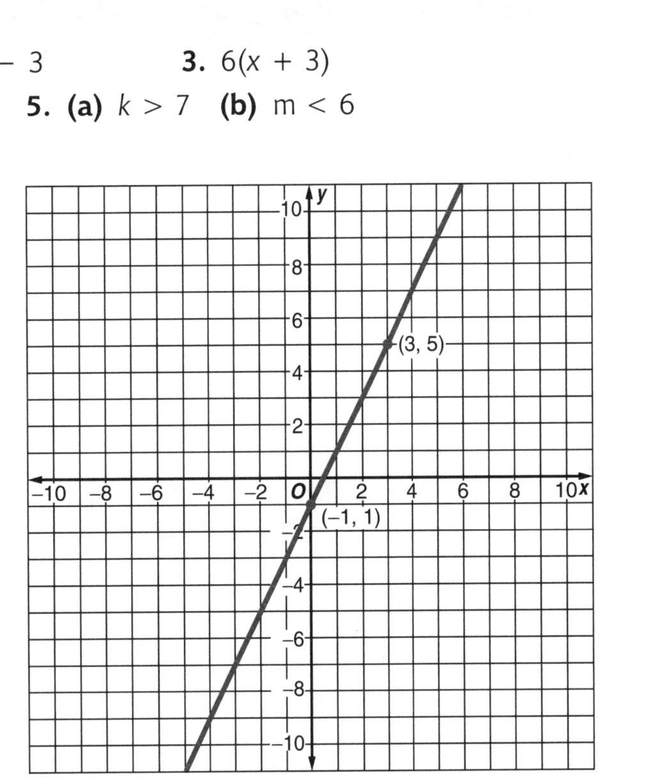

Outcome 3

7. 34.7 cm **8.** 30.7°

Outcome 4

9. (a) 380 000 000 000 **(b)** 0.000021

10. (a) 6.9×10^9 **(b)** 1.5×10^{-4}

11. 2.058×10^6 cm

Unit 4 Answers

Outcome 1

1. (a) £99.00 **(b)** Gross Pay: £539.00 Total Deductions: £128.92 Net Pay: £410.08

2. (a) £44.43 **(b)** £533.16

Outcome 2

3. (a) 2 **(b)** 164 miles (via Fort William and Aviemore)

4. £17

5. (a) (b)

F2 = C2 * D2

	A	B	C	D	E
1		Day	No. of Hours	Cost per Hour	Daily Total
2		Monday	3	£7.50	
3		Tuesday	5	£5.30	
4		Wednesday	2	£8.00	
5		Thursday	1	£4.80	

Sheet 1 Sheet 2 Sheet 3

(c) £22.50 **(d)** =SUM(E2:E5)

6. 100° **7. (a)** (scale drawing) **(b)** around 47 m **8.** 204 cm^2

9. A and C **10.** A + D

Answers to Practice Exam Questions

Unit 1 Answers

1. **(a)** 1.652 **(b)** 7800 **(c)** 2.82

2. £97

3. £57.20

4. **(a)** £582 **(b)** £379.20

5. 250 g

6. £63.38

7. 12.5%

8. £216

9. £50.52

10. 1.5 m

11. £23474.88

12. €1122

Unit 2 Answers

1. 9 hours 50 mins

2. **(a)** 36, 16, 5; total= 97; mean = 1.94; **(b)** $\frac{1}{2}$

3. **(a)** A(1, 3), B(3, 0) **(b)** on the *y*-axis 2 units below the origin **(c)** D(−2, 1)

4. **(a)** £3 **(b)** £6 **(c)** On average the girls got more (£5 compared to £3 for the medians); more variation in the amounts for boys (£6 range compared to £3 for the girls)

5. **(a)** 34 mph **(b)** 24 mph

6. 20

7. 10.1 m

8. **(a)** 6 **(b)** 21

9. Pile 1: $\frac{1}{3}$ Pile 2: $\frac{4}{9}$ so pile 2

10. **(a)** Positive **(b)** through (0, 0) and (100, 100) **(c)** around 85%

11. 2.9 m

12. **(a)** £207500 **(b)** £125000
(c) Median (unaffected by £650000)

Unit 3 Answers

1. $x = 4$

2. 0.002 86

3. **(a)** −8, −2, 4, 10 **3 (b)** see graph

4. −1

5. **(a)** $18m + 4$
(b) $4(2x+3)$

6. $x < 9$

7. 52.3 cm^2

8. C = 12

9. $x > 9$

10. 12.5

11. **(a)** 0, 1, 2, 3 **(b)** see graph

12. 3.2×10^6

13. $n = 11$

14. **(a)** $8x + 2$
(b) $4(6 - 5n)$

15. 4.9 m

16. 16

17. 2.7×10^{-2} sec.

18. **(a)** $3m - 7k$ **(b)** $8(2n - 3)$

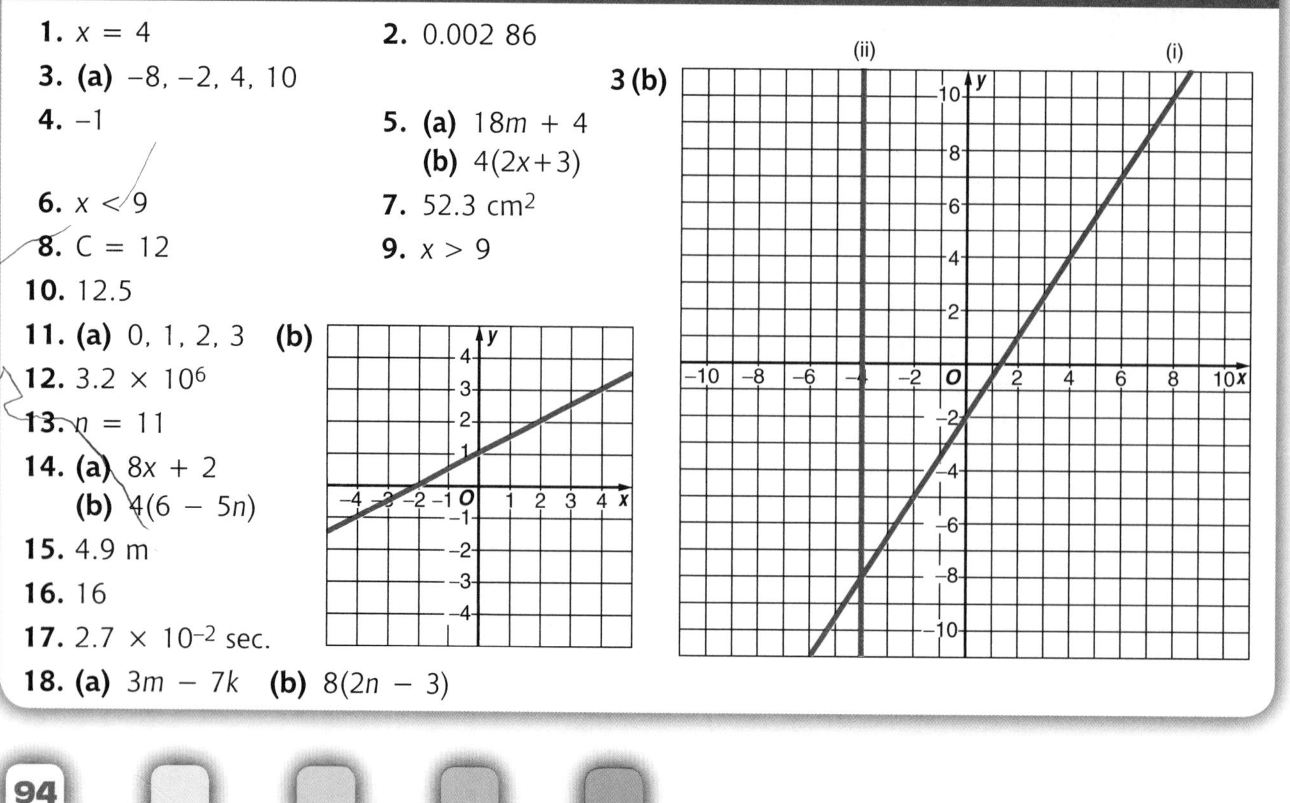

Unit 4 Answers

1. **(a)** £72.90 **(b)** £48.96

2. **(a)** 7.50 **(b)** = *B6* * *C6*

3. **(a)** 4 **(b)** 122 miles

4. **(a)** triangular prism
(b) 36 cm^2

5. **(a)** 1.1 kg
(b) Median weight is 2.2 kg for those that survived compared to 1.6 kg for those that died. This is 0.6 kg greater. The Range for those that survived is 2.5 kg, greater than 1.7 kg for those that died.

6. £3000

7. **(a)** 1 cm: 5km
(b) (Scale diagram)
(c) around 40 km

8. Overtime Rate: £7.80 Amount for overtime: £54.60
Gross Pay: £231.40 Total deductions: £74.50
Net Pay: £156.90

9. 4.5 m^2

10.

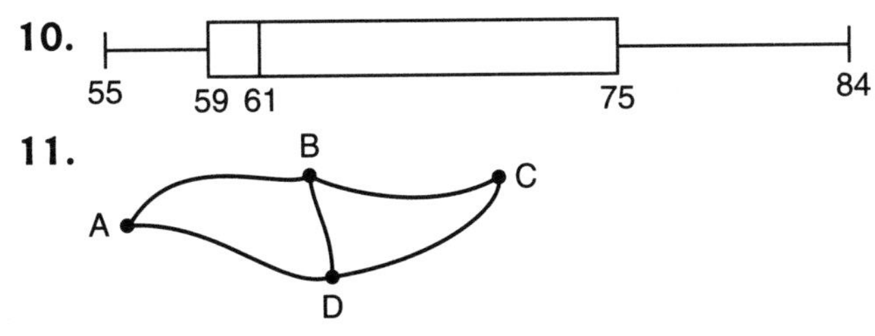

11.

Index

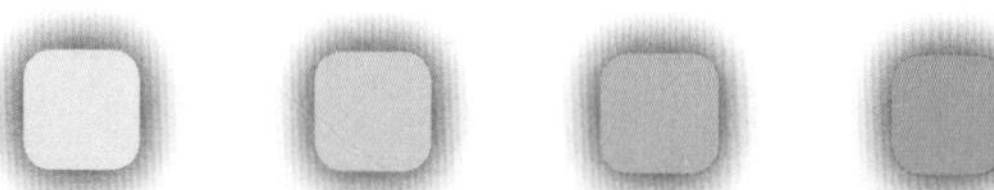

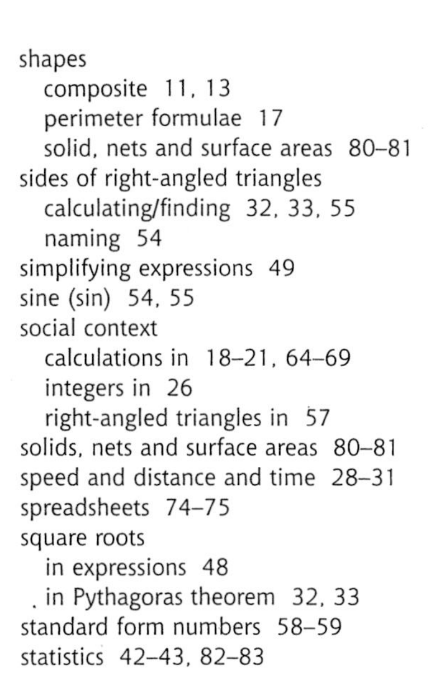